MW00712468

LOGIC

magazine

12

commons

winter 2020

LOGIC

ISSUE 12: Commons

EDITORIAL

Managing Editor	Alex Blasdel
Interviews Editor	Jen Kagan
Editors	Ben Tarnoff, Moira Weigel
Copyeditors	Victoria Gannon, Dave Stelfox
	Jacob Kahn

CREATIVE

Xiaowei R. Wang, Celine Nguyen

PRODUCTION

Jim Fingal, Christa Hartsock

/*

ISBN (print)	978-1-952550-06-5
ISBN (digital)	978-1-952550-07-2
ISSN	2573-4504

ELECTRONIC TELEGRAM

editors@logicmag.io

INTERNET

https://logicmag.io

Logic Magazine is published by the Logic Foundation, a California
nonprofit public benefit corporation organized for the specific
purpose of promoting education about technology.

*/

commons

**LOGIC
MAGAZINE**

Editorial

Portals

Features

Chatlogs

Assets

Contributors

Azad Amir-Ghassemi is a member of the Anti-Eviction Mapping Project's Los Angeles chapter and is a data analyst and urban planner by training.

Aaron Benanav is a researcher at Humboldt University of Berlin and the author of *Automation and the Future of Work*.

Sarah Brayne is an assistant professor of sociology at the University of Texas at Austin and the author of *Predict and Surveil: Data, Discretion, and the Future of Policing*.

Audrey Eschright is a writer, community organizer, and software developer based in Portland, Oregon.

Oscar H. Gandy Jr. is professor emeritus at the Annenberg School for Communication at the University of Pennsylvania.

Evan Malmgren is a research associate at Type Media Center who has written about power and communications technology for *The Baffler*, *Dissent*, *The Nation*, and others.

Erin McElroy is cofounder of the Anti-Eviction Mapping Project and is currently a postdoctoral researcher at the AI Now Institute at New York University, studying intersections of property, technology, race, and gentrification.

Charlton D. McIlwain is a professor of Media, Culture, and Communication at New York University, founder of the Center for Critical Race and Digital Studies, and the author of *Black Software: The Internet and Racial Justice, from the AfroNet to Black Lives Matter.*

Romi Ron Morrison is an interdisciplinary designer, artist, and researcher working across new media, black feminist praxis, and cultural geography.

Gavin Mueller is a lecturer in New Media and Digital Culture at the University of Amsterdam.

Nayantara Ranganathan is a researcher and lawyer writing on issues of technology and society.

Alex Hanna, Emily Denton, Andy Smart, and **Hilary Nicole** are researchers at Google. **Razvan Amironesei** is a research fellow at the University of San Francisco.

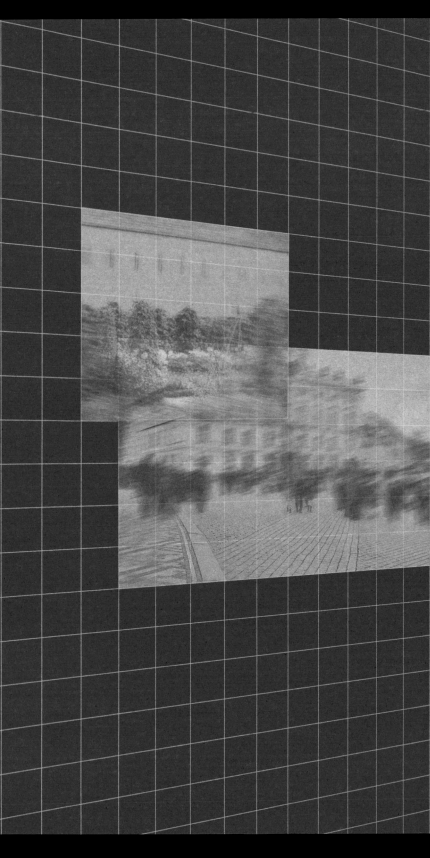

Singular Plural

1/

The internet began with the dream of a common language. The vision was a network of networks, bound together by a protocol that let a global community of computers speak to one another—an Esperanto, but for machines.

This dream expressed distinct and sometimes directly competing desires. It was built on wartime sciences of command and control, yet it also contained a communalizing impulse. On the early internet, everything was open source. Open standards prevailed over proprietary ones. Then, after the end of the Cold War, the US government gave the internet to the private sector.

The paradox followed that as the internet became truly commonplace, widely popularized, it gradually lost its openness: it became a set of walled gardens, dominated by the so-called platforms. The mediation of billions of lives through infrastructures owned by a handful of companies has not made us more free, even when their services are.

Neither has it made us more alike. As the internet has become more massified, it has created more differentiation: producing filter bubbles, epistemic crises, and ever finer demographic segmentation—not to mention amplifying inequalities between those who own a piece of the platforms and those who scramble to put together a living on them; those they give the God's-eye view, and those they're used to oversee. Despite the expectations of early US cyberutopians—or, you might say, imperialists—the global spread of the internet has not brought American values everywhere. On the contrary, there are strong tendencies toward fragmentation, from the General Data Protection Regulation in the European Union and calls for data sovereignty in South Asia and Latin America, to the billion users behind the Great Firewall of China. Even as networked technologies have become increasingly universal, it is debatable whether they are "common," in the deep sense of that word.

2/

Commons is an unusual kind of noun. Singular plural. The article it takes is: *The*. Historically, "the commons" has referred to the various kinds of resources that members of a community might share. Within village settings, for example, the commons consisted of a central green area where villagers grazed livestock, or a forest where a kind nobleman let them hunt, fish, and gather berries.

Elinor Ostrom, the economist who won the Nobel Prize for refuting the popular thesis of a "tragedy of the commons"—that is, the idea that any freely shared resource would inevitably be abused and deteriorate—said that a commons consists, at a minimum, of:

1. *Common goods*—Those fruits of nature and society that everyone needs to survive and thrive, including our atmosphere, oceans and forests, biodiversity, all species of life, natural systems, and minerals; our food, water, energy, and art; culture, technology, healthcare, and spiritual resources; and, also, news media, and the trade and finance systems we use.

2. *Commoners*—Groups of people who share these resources.

3. *Commoning*—Inclusive, participatory, and transparent forms of decision-making and rules governing access to, and benefit from, these common resources.

Ostrom also specified that commons came with boundaries. A commoning process, to include some, had to exclude others. What is needed, who needs it, and how to claim it are hotly contested political questions in our moment—particularly in the midst of a global pandemic. Who will ensure that the answers to these questions are found fairly?

This issue examines the theme of "commons" from all three of Ostrom's angles and more. Our authors investigate how large quantities of data have been collected and connected—not only by nation states, as the privacy advocates of the early internet feared, but by corporations, some of which sell their services back to the very government entities from which they were supposed to shield us. The fact that advertising, or attention capture, became the default business model of the internet is one reason for the situation in which we find ourselves. But, as this issue demonstrates, there are others. In the 2010s, advances in machine learning created powerful, and lucrative, incentives for companies to begin gathering as much data as they possibly could.

Alongside the companies that gather data, there are newly powerful companies that build the tools for organizing, processing, accessing, and visualizing it—companies that don't take in the traces of our common life but set the terms on which it is sorted and seen. The scraping of publicly available photos, for instance, and their subsequent labeling by low-paid human workers, served to train computer vision algorithms that Palantir can now use to help police departments cast a digital dragnet across entire populations.

What might have once looked like a *transgression* of the public-private boundary starts to look more like its *transformation*. But data can also be put to different purposes. Across the country, anti-eviction activists are using digital tools to extract information once held exclusively by corporate landlords and police departments, and put it into the hands of the tenant organizers who need it.

Our authors and interviewees also investigate who is setting the terms of our world-system: the lingua franca that our machines use to speak to one another, and we use to speak through them; the standards that govern infrastructures on which more and more of us depend. The past is a source of lessons, alternative visions, and practices that might help us thread the gap from present conditions to a livable future. "Freedom quilting" was a form of computation that was also a form of care work, and not only that.

Pieces in this issue ask how to organize a fairer system of algorithmic distribution, and what a more public, less commodified internet might look like. They explore why diversity initiatives have failed, and might have been designed to—and why opposing racism might require a radical transformation of the business model, not just new inputs to it.

We are writing from strange times. The luckiest among us have spent months in digitally mediated isolation. For others, the closing year has been a time of hunger, illness, and intensifying hopelessness. All of our lives are increasingly managed by algorithms that target and personalize. Then again, we have all been compelled to think with new keenness about our points of proximity. How intimate, to worry about breathing in air that a stranger in the supermarket has breathed.

In Latin, the phrase *locus communis* literally means "common place." Figuratively, it means a familiar topic of conversation. The "commonplace," in this sense, is not pejorative, like a cliché. It is a set of shared assumptions, a ground on which a diverse set of actors can meet. We hope this issue contributes to the making of that ground, to the construction of a space for us to think together about how, in whatever comes next, connected machines might help us create new collectivities and possibilities. ⁓

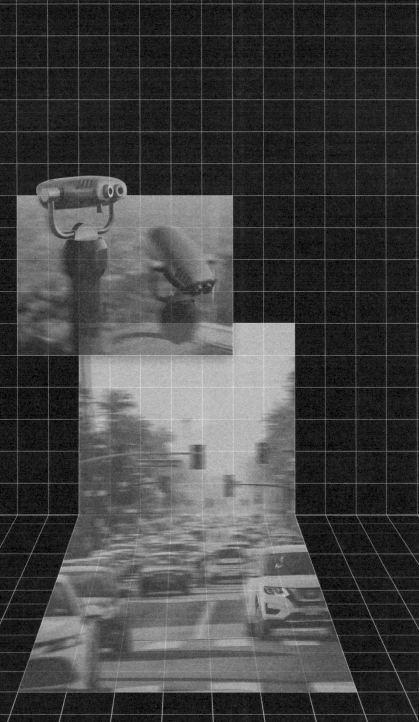

Enter the Dragnet

by Sarah Brayne

The LAPD is using Palantir to make everyone a target.

The Los Angeles Police Department's Real-Time Analysis and Critical Response Division (RACR) is housed in a hulking, institutional-gray building about a mile north of downtown. Its only marking is its address: 500. In the fall of 2013, I had an early morning meeting there with Doug, a "forward-deployed" engineer from Palantir Technologies, which builds and operates one of the premier platforms for compiling and analyzing massive and disparate data used by law enforcement and intelligence agencies. He was one of about eighty people I interviewed over the course of five years to understand how the LAPD uses Palantir and big data.

Palantir's goal is to create a single data environment, or "full data ecosystem," that integrates hundreds of millions of data points into a single search. Before Palantir, officers and analysts conducted mostly one-off searches in siloed systems: one to look up a rap sheet, another to search a license plate, another to pull up field interview cards, and more still to search for traffic citations, access the gang system, and so on. Seeing the data all together in Palantir is its own kind of data.

Palantir's clients—federal agencies such as the CIA, FBI, Immigration and Customs Enforcement, and the Department of Homeland Security; local law enforcement agencies such as the LAPD and New York Police Department; and commercial customers such as JPMorgan Chase—need training in order to learn how to use the platform, and they need a point person to answer their questions and challenges. That's where engineers like Doug come in.

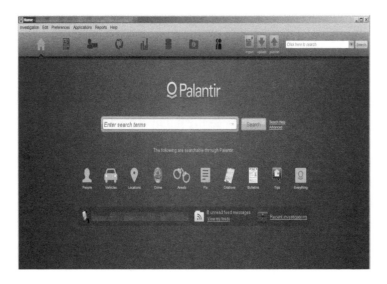

FIGURE 1 Palantir homepage.
SOURCE Los Angeles Police Department

In a training room at the RACR, I watched as Doug logged in to Palantir Gotham, the company's government intelligence platform, and pulled up the homepage (Figure 1). For him, it was a banal moment, but I'd been eagerly anticipating this sight: there is virtually no public research available on Palantir, and media portrayals are frustratingly vague.

Doug started running through some of the different ways the platform could be used. He patiently explained what he was doing as he queried, clicked, zoomed, narrowed, and filtered:

> So now, imagine a robbery detective who says, "Hey, you know what, I have a male, average build, black four-door sedan." Like, they would [previously be able to] do nothing with that, right? So, we can do that.

> Let's go take a look at vehicles that are in the system... There are 140 million records in this system ... we know it's a Toyota, maybe a Hyundai, right? Or a Lexus... So let's say we think it's one of those types of vehicles, right? And that got us then to 2 million [vehicles]. And if we were to go look at, say, a color ... we know it was black. Maybe it was blue, 'cause it could have been blue. It could be dark green... And we know it was a four-door.

> Do you see what's happening over here? In five hops, they're able to get down to 160,000. Now they're still not going to look at 160,000 vehicles. We didn't get into model and year, but we could do that, and we could chart it, which makes it easy... So now I could say, I think it was between 2002 to 2005, drill down, now we're 23,000. Now it gets pretty manageable.

> So now let's flip over and let's look at the people that are connected to these vehicles. And I know I'm looking for a male. And I'll just do one of them.

And I know that, like, let's say he was pretty short. And he was on the heavier side. Brick house. We just got down to thirteen objects, thirteen people. And you could say, "Okay, well, now let me take a look at—all thirteen have driver's license numbers." So now we've narrowed it down to thirteen potential people, and they could take these thirteen objects and go to the DMV and pull their DMV photos and go to the witness or victim and say, "Here you go."

In less than a minute, using partial information, Doug was able to narrow a search from 140 million records to thirteen. He went on to show me how to look up which of the thirteen had any citations or arrests, the LAPD divisions in which they received their citations or were arrested, and identify one person who had been cited in the same division in which the robbery occurred. If the person ended up not being the person who committed the robbery, officers could save this search formula and keep running it in the coming days, in case any new data came in.

I asked what happens when the system gives a false positive. What happens to the wrong suspect? Doug said bluntly, "I don't know."

One Person's Click Is Palantir's Clue

We all leave hundreds of digital traces—clues, should it come to that—every day. When we use our cell phone, run an internet search, or buy something with a credit card, we leave a digital trace. Rapidly proliferating automatic data-collection sensors record and save those digital traces, and make dragnet surveillance—the collection and analysis of information on everyone, rather than only people under suspicion—possible at an unprecedented scale. Individuals with no direct police contact are now included in law enforcement systems, and police now collect and use information gleaned from institutions, like credit-rating agencies, that are typically not associated with crime control.

But data—particularly large, diverse sets of data—are relatively useless on their own. You need a good platform to sort through them. And Palantir is excellent at processing, sorting, and analyzing data. With the right platform, searches that used to take hours, days, or even weeks may now take only a few seconds. Dragnet surveillance—and the data it produces—can be incredibly useful for law enforcement to solve crimes. As one officer explained, after any crime, "the first thing you're gonna do, always, is check the digital footprint."

People working in information technology have a vested interest in making the case that information technology is a crucial component of law enforcement. But no matter how you quantify it—through increased federal and within-department funding for data-intensive policing, the proliferation of law enforcement contracts with tech companies, the increase in tech-training sessions for police, or the rising number of data points the police access daily—data analytics are central to law enforcement operations today.

The Southern California fusion center Joint Regional Intelligence Center (JRIC), a multiagency, multidisciplinary surveillance organization, started using Palantir in 2009 to connect and analyze Suspicious Activity Reports (SARs). At the time, it was the largest law enforcement deployment of this software anywhere in the world. The LAPD, Long Beach Police Department, and LA City Fire Department soon adopted the platform, and there were over 1,300 trained Palantir users in the region by 2014. More users are onboarded every month. A sergeant named Michaels, who coordinates some of the training sessions at JRIC, claims "they catch bad guys during every training class."

The LAPD's arrest records and field interview cards—small, double-sided index cards that officers fill out with key information about people they interact with in the field—were the first data sources integrated into Palantir. Both are geocoded, meaning you can plot where these police stops and arrests occurred. Palantir does not own the data the LAPD uses, but rather provides an interface that makes it possible to link data points across previously separate systems. Users can plot data on maps, as a network, as a time wheel, or as a bar graph with a timeline of phone calls and financial transactions, for example. The platform even allows users to organize structured and unstructured data content such as emails, PDFs, and photos through tagging.

Because one of Palantir's biggest selling points is the ease with which new, external data sources can be incorporated into the platform, its coverage grows every day. LAPD data, data collected by other government agencies, and external data, including privately collected data accessed through licensing agreements with data brokers, are among at least nineteen databases feeding Palantir at JRIC. The data come from a broad range of sources, including field interview cards, automatic license plate readings, a sex offender registry, county jail records (including phone calls, visitor logs, and cellblock movements), and foreclosure data.

Though there was a lot of uncertainty among my interviewees about exactly what data were in the databases they were accessing, a pair of civilian employees mentioned the use of LexisNexis's public records database Accurint and speculated that it contained documents like utility bills and credit card information. Indeed, LexisNexis has over 84 billion public records from 10,000 diverse data sources, including 330 million unique cell phone numbers, 1.5 billion bankruptcy records, 77 million business contract records, 11.3 billion name and address combinations, 6.6 billion motor vehicle registrations, and 6.5 billion personal property records.

"One of Palantir's biggest selling points is the ease with which new data sources can be incorporated. Its coverage grows every day."

The process of labeling and linking objects and entities like persons, phone numbers, and documents makes it possible to plot data on maps and graphs that let users see data in context and make new connections. It can also make it easier to see what crucial data might be missing or what sorts of data might be useful for law enforcement to begin collecting. Whereas one piece of information may not be a useful source of intelligence on its own, Doug explained, the "sum of all information can build out what is needed."

The Danger Imperative

Most sworn officers and civilian staffers, including crime analysts, who use Palantir Gotham use it for simple queries (what Palantir calls "drilling down using 'object explorer'"). Users can search for anything from license plates to phone numbers to demographic characteristics, and a vast web of information will be returned. One officer described the process:

> *You could run an address in Palantir, and it's going to give you all the events that took place at that address and everyone who's associated to those events... So if it's a knucklehead location where a lot of things are happening there, you're gonna get people documented on there one way or the other... Either field interview cards, or they're on crime reports, whatever... Otherwise you could search*

all of the records within [LexisNexis's] Accurint ... and see who's living with who a lot of times.

It's not quite so simple as "run a query, get a list of suspects." Notice, for example, how the officer says "knucklehead location"—he means that you can query an address, and if it's been listed as the address for many people or their car registrations, if it's been the site of multiple calls for service, or it's otherwise connected across the databases, a Palantir Gotham search is going to return a tangle of information. Some of it will be useful, some of it won't, but the presumption is that if criminal activity is going on at a location, *someone* or *something* will be in Palantir.

Another employee at Palantir demonstrated how the platform can be used for retroactive investigative purposes: Law enforcement had a name of someone they thought was involved in trafficking. They ran a property search, which yielded the person of interest's address and date of birth. Then they ran a search for common addresses (whether there are any other people in the system associated with the same address). One turned out to be a sibling of the initial person of interest, which sent investigators searching again, this time coming up with a police report for operating a vehicle without a license. They also searched the address of a third sibling, who lived at a different address. A radius search revealed several tips concerning this same house: one neighbor had called in to report a loud argument, and another reported that a suspicious number of cars was stopping at the house.

With this information, the police were able to set up in-person surveillance and subpoena phone records, which were run through Palantir's "time wheel" function to identify temporal patterns. Modeling revealed phone calls to one or two phone numbers at the same time each week; using those phone numbers, police got a new database hit. They found a name and a police report and identified their suspect.

In another instance, I saw a user search for a car using just a partial license plate. They entered "67" and accessed all of the crime reports, traffic citations, field interview cards, automatic license plate readings, names, addresses, and border crossings associated with cars whose license plate contained these numbers in this order.

Advanced analytic tools on the platform include geo-temporal and topical analysis, each of which can be visualized differently. For example, users can plot (geo-analysis) all the types of crime they are interested in (topical analysis) during a given period of time (temporal analysis). Users can visualize the data on a map or along a chronological axis, as well as conduct secondary and tertiary analyses in which they analyze the results by, for example, modus operandi (e.g., using a bolt cutter) or proximity of robberies to a parolee's residence.

Another way to use the analytic suite is to paint a detailed picture of the population of interest in an area. One officer explained this:

> The big thing that Palantir offers is a mapping system. So, you could draw out a section of [his division] and say, "Okay, give me the parolees that live in this area that are known for stealing cars" or whatever [is] your problem... It's going to map out that information for you ... give you their employment data, what their conditions are, who they're staying with, photos of their tattoos, and, of course, their mugshot. [And it will show] if that report has [a] sex offender or has a violent crime offender or has a gang offender. Some are in GPS, so they have the ankle bracelet, and ... we have a separate GPS tracker for that.

A Palantir software engineer spoke of the gang unit monitoring entire networks of people: "Huge, huge network. They're going to maintain this whole entire network and all the information about it within Palantir."

"This is one of the most transformative features of big data: the creep of criminal justice surveillance into other institutions."

Palantir, one sergeant explained, is also an "operational game changer": it gives him the data he needs to protect his officers' safety by, for instance, locking down a neighborhood and positioning an airship overhead while law enforcement conducts a search. Of course, this situational awareness made possible by Palantir can ratchet up officers' sense of danger and escalate an already tense situation. Such platforms provide an unprecedented number of data points supporting the "danger imperative," or the cultural frame officers are socialized into, which encourages them to believe that they may face lethal violence at a moment's notice.

Criminal Justice Creep

New data sources are incorporated into Palantir regularly. One captain commented:

> I'm so happy with how big Palantir got... I mean it's just every time I see the entry screen where you log on there's another icon about another database that's been added ... they just went out and found some public data

on foreclosures, dragged it in, and now they're mapping it where it would be relative to our crime data and stuff.

Another interagency data integration effort is LA County's Enterprise Master Person Index (LA EMPI) initiative. If established, LA EMPI would create a single view of an individual across all government systems and agencies: all of their interactions with law enforcement, social services, health services, mental health services, and child and family services would be in one place under a single unique ID. Although the explicit motivation behind the EMPI initiative is to improve service delivery, such initiatives extend the governance and social control capacities of the criminal justice system into other institutions.

This is one of the most transformative features of the big data landscape: the creep of criminal justice surveillance into other, non–criminal justice institutions. I encountered many examples of law enforcement using external data originally collected for non–criminal justice purposes, including LexisNexis, but also TransUnion's TLOxp (which contains one hundred billion public and proprietary data points, including social security numbers, employment records, and address records); databases for repossession and collection agencies; social media, foreclosure, and electronic toll pass data; and address and usage information from utility bills.

Respondents added that they were working on integrating hospital, pay-parking lot, and university camera feeds, as well as rebate data, pizza chain customer lists, and so on. One interviewee in the LAPD's Information Technology Division said they had their eye on consumer data: "Other stuff, shopping data. You can buy it, you know, certainly other vendors are. So why not?" In some instances, it is simply easier for law enforcement to purchase privately collected data than to rely on in-house data, partly because there are fewer protections and less oversight over private sector surveillance and data collection.

Another of the most substantively important shifts that have accompanied the rise of big data policing is the shift from query-based systems to alert-based systems. By "query-based systems," I mean those databases that operate in response to a user query, such as when an officer runs your license plate during a traffic stop. In alert-based systems, by contrast, users receive real-time notifications when certain variables or configurations of variables become present in the data. High-frequency data collection makes alert-based systems possible, and that carries enormous implications for the relational structure of surveillance.

Imagine an officer wants to know about any warrants issued for residents of a specific neighborhood. In a query-based system, they would need to set up specific searches, and most of those would be useful only well after the warrant had been issued. All of the millions of warrants in LA county are geocoded and can be translated into object representations spatially, temporally, and topically in Palantir. Through tagging, users can add every known association of a warrant to people, vehicles, addresses, phone numbers, documents, incidents, citations, calls for service, automatic license plate readings, field interviews, and the like. All that information is cross-referenced in Palantir. Then, using a mechanism in Palantir that's similar to an RSS feed, an officer can set up automatic notifications for warrants or events involving specific individuals (or even descriptions of individuals), addresses, or cars to ping their cell phone.

For example, an alert can be set up by putting a geofence around a given area and requesting an alert every time a new warrant is issued within that perimeter. One sergeant had an email alert set up in this way, and could even get the alert while he was out on patrol. "Court-issued warrant, ding!" As soon as he got the notification, he says, he was able to track down and arrest the suspect. Previously, the process was far slower. "Now," he explained excitedly, "you draw a box in Palantir and go about your business. Ding!"

A civilian employee described a similar approach using automated license plate readings: "If you have an automated license reader, you can flag a plate or a partial plate and you could attach it to your email. And if it ever comes up, it will send you an email saying, 'Hey, this partial plate or this vehicle, there was a hit last night. Here is the information.'"

Becoming Carmen Sandiego

Law enforcement databases have long recorded who has been arrested or convicted of crimes. Today, they also include information on people who have been stopped, as evidenced by the proliferation of stop-and-frisk databases. The real surprise may be that as new data sensors and analytic platforms are incorporated into law enforcement operations, the police increasingly utilize data on individuals who have not had any police contact at all.

The automatic license plate reader (ALPR) is perhaps the clearest example of a low-threshold "trigger mechanism," lowering the bar for criteria that justifies inclusion in databases. ALPRs are quintessential dragnet surveillance tools—they take readings on everyone, not just people under suspicion. Their data come in the form of two photos—one of the license plate and one of the car, along with the time, date, and geo-coordinates attached to those photos, as read by ALPR cameras mounted on the tops of police cars and static cameras at intersections and

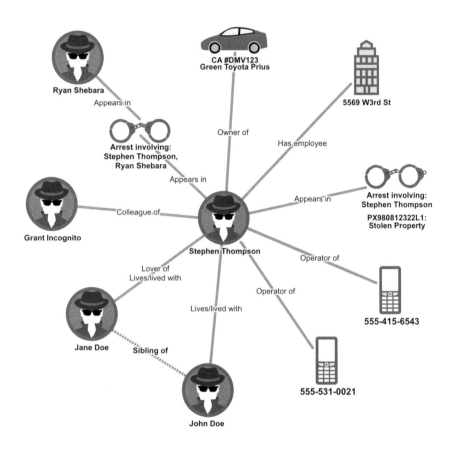

FIGURE 2 Network in Palantir.

SOURCE Palantir Technologies; diagram by David Hallangen.

other locations. ALPR data collected by law enforcement can be supplemented with privately collected ALPRs, such as those used by repossession agents. Just this one relatively simple technological tool makes everyday mass surveillance possible on an almost unimaginable scale.

In addition to ALPRs, there are all sorts of low-threshold trigger mechanisms being leveraged by the LAPD. Much of the data is what's being called "collateral data collection," and it is a passive, pervasive way people are being caught up in the surveillance state. Figure 2 is a de-identified notional representation, based on a real network diagram I obtained from the LAPD.

The Carmen Sandiego–looking figure in the middle, "Stephen Thompson," is a person with direct police contact. Radiating outward, we see all the entities he is related to, including people, cars, addresses, and cell phones. Each line indicates the type of connection (e.g., sibling, lover, co-arrestee, vehicle registrant).

> "*Digital traces can be knit together so that circumstantial evidence looks like a comprehensive picture.*"

To be in what I call the "secondary surveillance network," radiating out from the person of interest, individuals do not need to have direct law enforcement contact; they simply need a connection to the central person of interest. And once they are in this system, these individuals can be autotracked, meaning officers can receive real-time alerts should they come into contact with the police or other government agencies.

When many streams of information flow together, they form a "data double," which can be a powerful tool in the hands of law enforcement. As a member of legal counsel at Palantir explained, digital traces can be knit together so that circumstantial evidence looks like a comprehensive picture: there is "usually not one smoking gun document, but we're able to build up a sequence of events prosecutors might not previously have been able to do ... [we can integrate] data in a single ontology to rapidly connect illicit actors and depict a coherent scheme." This reconstruction may be invisible to civilians — and to their lawyers, if they end up being charged.

But indiscriminate data collection is not the inevitable outcome of technological advancement. Mass surveillance is not the "natural" result of mass digitization. Instead, what we allow to proliferate and become the objects of massive data-collection efforts are choices that reflect the social and political positions of the subjects and subject matter that we feel comfortable surveilling.

As a counterpoint, consider guns in the United States: we do not permit the mass tracking of guns. There is no federal gun registry, and the National Instant Criminal Background Check System is *required* by law to destroy the audit logs of background checks that go through its system within ninety days. We certainly have the technology to track guns, and we could easily leverage existing technology to do more tracking, but gun owners are powerful political subjects. They have the resources to assert that their guns should not be tracked.

Police officers, too, have routinely invoked their authority and legitimacy to undermine attempts to surveil *their* work lives. They have the power to resist in ways that their more usual subjects, disproportionately low-income, minority folks with little political capital and no small amount of fear, cannot. In that way, too, dragnet surveillance serves to reinscribe inequality. ∿∿

Sarah Brayne is an assistant professor of sociology at the University of Texas at Austin and the author of *Predict and Surveil: Data, Discretion, and the Future of Policing*.

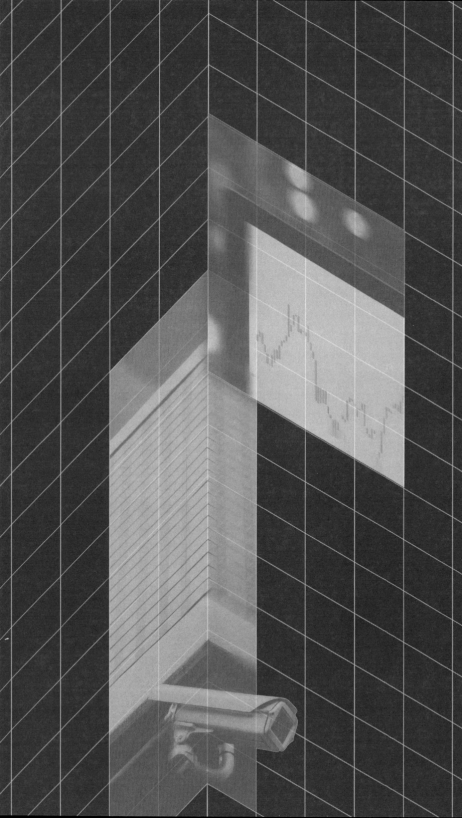

Panopticons and Leviathans

Oscar H. Gandy Jr. on Algorithmic Life

The story of surveillance capitalism is older than Google. Even before the internet became a mass medium, private firms were using the computerized collection and processing of data in order to classify and manipulate people. The scholar Oscar H. Gandy Jr. has been studying this phenomenon from the start—and sounding the alarm about the dangers it represents. An expert on the political economy of personal information, he was a professor at the Annenberg School for Communication at the University of Pennsylvania for nearly twenty years. He is now retired and lives in Arizona, but continues to write and give talks. We spoke to him about how exactly he predicted our present, and whether there's any hope of taming the algorithmic Leviathan.

––––––––

Back in 1993, you published a book called *The Panoptic Sort*. In it, you described how the computerized collection and processing of personal information was creating an all-encompassing surveillance regime that, by sorting

people into categories and classes, shaped their lives by controlling their access to goods and services.

When I read it last year, I found it incredibly prescient. In the early 1990s, the internet hadn't yet gone mainstream, and digitization wasn't nearly as sophisticated or comprehensive a process as it is today. Yet you identified an emerging phenomenon that, almost thirty years later, has become the central organizing principle of our digital lives. How did you see the contours of this trend so early? How do you see it now?

At the time, the book was supposed to be a challenge to the way that most policy scholars were thinking about privacy. For them, government was the major focus of concern—they were worried about *governmental* invasions of privacy. I wanted to shift the focus to corporate surveillance: the gathering of information by businesses in order to produce market segments and targets. In retrospect, that turned out to be an appropriate focus.

The mid-1990s was also the era when techno-libertarianism—as developed by Kevin Kelly, John Perry Barlow, *Wired*, and others—was gaining influence. The state was seen as the principal enemy. And in fairness, they had a point—the Communications Decency Act of 1996 did represent an attempt by the government to censor the internet. But the techno-libertarian approach could also downplay or even completely ignore the threat posed by corporate power.

Right. But of course, a lot has also changed since then. I was writing about the kind of data gathering and analysis being done by researchers at firms, or researchers working as consultants to firms. There has been a really substantial shift in the nature of that work, because it's now being done by algorithms. Algorithms are now taking on, or being assigned, greater responsibility for the kinds of questions that are being asked and the kinds of relationships that are being explored. That's a major shift, and one that I've certainly been paying attention to. And I'm hoping that the rest of the world is paying attention as well, because there are going to be new consequences with a new actor.

> "*Government was the major focus of concern... I wanted to shift the focus to corporate surveillance.*"

I used to struggle with some of my graduate students in talking about algorithmic systems as "actors" in this regard. But we've got to understand them as actors in order to be able to assign responsibility—whether it's through legal means or some other kind of tool. These systems are doing assessments, making classifications, generating predictions, and designing interactions in order to influence our behavior.

How do you hold an algorithmic actor responsible for its actions?

First we have to understand how the law is limited by its historic focus on the individual. It is structured around the idea that privacy invasions are attacks on *individual* liberty. But algorithmic processing is about groups.

There are certain groups that federal law has designated "protected classes." For instance, employers cannot discriminate on the basis of race, color, national origin, religion, sex, age, or disability. But we're really behind the eight ball in terms of *algorithmically-defined* groups. These groups have limited political capability, because their members don't understand the nature or even the identity of the groups to which they are being assigned—even though their membership in these groups serves as the basis on which they are discriminated against by commercial and state actors that have an interest in manipulating them.

In other words, we all have identities that we're not aware of—identities that a computer has constructed in order to make it easier for a company to sell us things, or for the state to lock us up. But because we don't have access to what these identities are, or knowledge of how they're made, it's hard for us to organize around them politically. This is in contrast to the identities that make up federally protected classes, which reflect the achievements of struggles by members of those groups who composed themselves into a political bloc.

"Individuals are placed within these new categories, these new identities, because these identities matter to the actors who are relying upon algorithmic systems in order to influence the behavior that matters to them."

There is a whole host of technologies that have to do with identifying individuals as members of groups in order to make predictions, in order to estimate things like value or risk. For instance, a firm might calculate insurance rates based on where you live, or on the characteristics of the people within your neighborhood, and the estimation of risk associated with those factors. They are legally prohibited from using race to calculate those rates, but they can still use proxies for race, intentionally or not.

If a bank says it won't lend to someone because they're Black, it's illegal. But if a bank uses an algorithmic system that ingests a bunch of data and performs an analysis that in effect infers that the prospective borrower is Black—say, they live in a majority-Black zipcode—and then denies them on that basis, they can get away with it.

Correct. Discrimination is continuing on the basis of race, gender, and other categories. This kind of discrimination—against groups whose members self-identify and therefore relate to each other and mobilize politically on the basis of that shared identity—is very important. But what I'm trying to get us to pay attention to is the other groups that we have been assigned membership, and through which we experience discrimination, but which we *know nothing about*.

This relates to a distinction you draw in your work between "identity" and "identification." Identity, you write, "is primarily the result of personal reflection and assessment, something closely associated with individual autonomy." Identification, on the other hand, is "almost entirely the product of the influence and determination of others." Identification is a social process, in other words, mediated by various digital technologies. And once an individual is identified, they can be classified into a group, and subjected to statistical analysis.

And discriminated against, and manipulated with nudges in order to shape their behavior.

Individuals are placed within these new categories, these new identities, because these identities matter to the actors who are relying upon algorithmic systems in order to influence the behavior that matters to them. That matters to them as capitalists, perhaps, or that matters to them as governors and mayors and others in the political realm. The actors are different, and they have related but different motivational factors. But they are all making decisions with the aid of algorithmic systems that identify people, and then direct manipulative communications towards them in order to influence their behavior. And, to push it one step further: to influence who you are, who you want to be, how you think you ought to change in order to become the kind of person you are being led to believe you should be.

How do these processes of identity and identification interact? On the one hand, there's clearly a tension between an identity that I'm conscious of and which is important to me, and an identity that I'm not aware of and is important primarily to the state or corporate actors who want to influence my behavior.

Yet as more of our lives is mediated by digital technology, these technologies also become the medium through which many of us come to know ourselves—where our identity actually undergoes formation. Social media comes to mind. Of course, social media platforms are major sites of what you call identification: software is observing our activity on these platforms in order to sort us into groups we know nothing about so that our attention can be better sold to advertisers. But people are also constructing their identities quite consciously through their interactions on these platforms, and those identities can in turn produce real political effects by triggering new waves of social mobilization.

Part of what we are seeing with social media is further diversification within all categories of identity. There are new kinds of identities that we may not even be able to articulate yet, but which are being reflected in new kinds of social movements, such as the movement in the months following the death of George Floyd. All of these white folks are out there engaging in active demonstration against anti-Black abuses by police. Something has clearly happened, in part through social media, which has led to new forms of identity emerging among white people.

So yes, we need to address the variety of identities that people are aware of, including the many new ones being created. But we also need to address those that we don't have names for yet, the ones that are being generated by algorithmic manipulation. The bottom line is that we are in the midst of a historical moment in which both identity and identification are undergoing dramatic change.

Risk Factors

We've been talking about algorithmic logic. But the focus of your most recent book, *Coming to Terms With Chance*, is on *actuarial* logic—specifically, "the actuarial logic that shapes the distribution of life chances in society." You tell a story about how society became obsessed with assessing and managing risk, and the role that probability and statistics have played in this shift. What does it mean to live in such a society, and how did we get there?

Let's start with the term "life chances." What are the chances for good things to happen to us? What are the chances for bad things to happen to us? And what shapes those life chances?

Increasingly, the decisions that influence our life chances are made on the basis of statistics. The probability that we're going to have a good future versus the probability that we're going to have a bad future is itself determined through practices of probabilistic analysis. And this has enabled the emergence of something I call "rational discrimination."

Rational discrimination is when discrimination becomes justified in terms of an assessment of risk that can be said to be rational. It is both a methodology and a way to make discrimination acceptable in the eyes of the law. It is rooted in the argument that it is justifiable to discriminate against people—including people who can be identified by race, gender, and a host of other attributes—where there is statistical evidence of risk.

This gets back to our conversation about using proxies for race to perpetuate racial discrimination without formally discriminating by race. In this case, the bank is not saying it won't give someone a loan because they're Black. It's saying that an algorithm told them that the individual has too high a risk of default, so they can't get a loan.

But then why are members of certain groups considered riskier than others? This is where we need to talk about "cumulative disadvantage." For example, some of these models make predictions on the basis of an individual's level of education. Well, we know the education system is highly unequal. Therefore, there is cumulative disadvantage as a result of the kinds of differences in education that people have, because those differences are then used to discriminate against them.

And it's not just education, of course. There are all sorts of factors that are subject to cumulative disadvantage. And these will continue to perpetuate discrimination unless there is a powerful actor that steps in and limits the use of certain factors in making predictions. Otherwise, the harms that are associated with cumulative disadvantage will just pile up.

How new is the practice of rational discrimination? I'm reminded of the redlining maps that government officials and banks developed in the 1930s to deny certain neighborhoods access to federally backed mortgages. These neighborhoods were predominantly Black and Latino, but the formal basis for excluding them was that they had a higher risk of default.

True enough. Look, I'm an old guy. I did statistics by hand. Statistics has been around for ages. The estimation of risk has been around for ages. And while discrimination on the basis of race may not have been based upon statistics at first, it soon was. But the nature of statistics has changed, and the nature of the technologies that use statistics have changed, in part through rapid developments in computation. That's what we've got to pay attention to, especially if we want to gain control over these systems.

> "*Increasingly, the decisions that influence our life chances are made on the basis of statistics.*"

These statistical systems infer the future from the past. But is this a reliable mechanism in our historical moment? As we're speaking, there are red skies over San Francisco. Extreme weather events are only going to increase as climate change gets worse. We're also clearly entering a new era of intensified social and political conflict. It seems likely that the next

few decades will be full of events that may be hard to predict by looking at the past. Does this create a vulnerability for these systems?

It's true that for these systems, knowledge of the future is based on knowledge of the past. But the past is getting to be very short. If you think about big data, if you think about the systems being used by Google and Facebook, decisions are being made continuously in response to new data. So forget the past. Seriously, forget the past. The data and the models are being altered daily. The most powerful of them are being altered moment to moment. So the past is not relevant anymore. Not really. Not meaningfully.

> "*I'm not ready to grant a single Leviathan that wisdom. I believe in the multiple. I believe in the differences among us.*"

What's the best path forward towards a better future? In the final chapter of *Coming to Terms With Chance*, you call for "a social movement to oppose expanded use of statistical techniques for the identification, classification, and evaluation of individuals in ways that contribute further to their comparative disadvantage." Might that offer a path forward?

It's lost. There's no chance. That's gone.

The future is the algorithm. The future is what Pascal König calls "the algorithmic Leviathan."

Tell me more about this algorithmic Leviathan.

In the Foucauldian panopticon, you have a central tower. The central tower has tremendous power because the population believes that observers within the tower are able to see what everyone is doing—even though there is no way to know for sure. The way the tower operates is that we learn the behaviors that are expected of us, and modify our behavior accordingly.

The Leviathan is similar. But there is no central tower. Rather, you have an algorithmic system that doesn't need to be located in a central place because we are now in a networked environment. It doesn't need to be in a place at all, it just has to be in the network. More specifically, it has to be in a position within the network where it has access to the data that has been gathered by all of the responsible elements within the network. Those responsible elements within the network have their own subsystems and their own sub-networks for gathering the information that matters for them. The Leviathan provides the control systems for this gathering and consumes the data that results from it.

In the panopticon, the central tower is feared. The Leviathan, however, is a trusted figure. A god-like figure that is trusted to act in our individual and collective interest.

But I don't want to trust. And I don't want the rest of us to trust. I don't trust systems. Systems are built by designers who work for corporations that have very specific ideas about how such systems should work.

Is there any hope then of taming the Leviathan?

One could imagine a community of algorithmic systems. A plethora. These systems would be designed to have a socially agreed-upon wisdom. I'm not ready to grant a single Leviathan that wisdom. I believe in the multiple. I believe in the differences among us. We can make systems that embody those differences: not quite Leviathans, but committed resources that stand in for the differences among us. That's about as close as I can get to envisioning a better future at the moment.

I'm not ready to write about it yet. I'm thinking about it. I'm working on it. If you look at my resume, you can see that I write books ten years at a time. But I don't think I have ten more years. I really don't. So if I'm able to make the Leviathan my next thing, or my last thing, that's good enough for me. ⌇

Lines of Sight

by Alex Hanna, Emily Denton, Razvan Amironesei,
Hilary Nicole, and Andrew Smart

You can't understand the AI systems that are transforming our
world without understanding the datasets they are built on.

———

On the night of March 18, 2018, Elaine Herzberg was walking
her bicycle across a dark desert road in Tempe, Arizona. After
crossing three lanes of a four-lane highway, a "self-driving"
Volvo SUV, traveling at thirty-eight miles per hour, struck her.
Thirty minutes later, she was dead. The SUV had been operated
by Uber, part of a fleet of self-driving car experiments operating
across the state. A report by the National Transportation and
Safety Board determined that the car's sensors had detected an
object in the road six seconds before the crash, but the software
"did not include a consideration for jaywalking pedestrians." In
the moments before the car hit Elaine, its AI software cycled
through several potential identifiers for her—including "bicycle,"
"vehicle," and "other"—but, ultimately, was not able to recognize
her as a pedestrian whose trajectory would be imminently in the
collision path of the vehicle.

How did this happen? The particular kind of AI at work in autonomous vehicles is called machine learning. Machine learning enables computers to "learn" certain tasks by analyzing data and extracting patterns from it. In the case of self-driving cars, the main task that the computer must learn is *how to see.* More specifically, it must learn how to perceive and meaningfully describe the visual world in a manner comparable to humans. This is the field of computer vision, and it encompasses a wide range of controversial and consequential applications, from facial recognition to drone strike targeting.

> *"Machine learning enables computers to 'learn' certain tasks by analyzing data and extracting patterns from it."*

Unlike in traditional software development, machine learning engineers do not write explicit rules that tell a computer exactly what to do. Rather, they enable a computer to "learn" what to do by discovering patterns in data. The information used for teaching computers is known as *training* data. Everything a machine learning model knows about the world comes from the data it is trained on. Say an engineer wants to build a system that predicts whether an image contains a cat or a dog. If their cat-detector model is trained only on cat images taken inside homes, the model will have a hard time recognizing cats in other contexts, such as in a yard. Machine learning engineers must constantly evaluate how well a computer has learned to perform a task, which will in turn help them tweak the code in order to make the computer learn better. In the case of computer vision, think of an optometrist evaluating how well you can see. Depending on what they find, you might get a new glasses prescription to help you see better.

To evaluate a model, engineers expose it to another type of data known as *testing* data. For the cat-detector model, the testing data might consist of both cats and other animals. The model would then be evaluated based on how many of the cats it correctly identified in the dataset. Testing data is critical to understanding how a machine learning system will operate once deployed in the world. However, the evaluation is always limited by the content and structure of the testing data. For example, if there are no images of outdoor cats within the testing data, a cat-detector model might do a really good job of recognizing all the cats in the testing data, but still do poorly if deployed in the real world, where cats might be found in all sorts of contexts. Similarly, evaluating Uber's self-driving AI on testing data that doesn't contain very many jaywalking pedestrians will not provide an accurate estimate of how the system will perform in a real-world situation when it encounters one.

Finally, a *benchmark dataset* is used to judge how well a computer has learned to perform a task. Benchmarks are special sets of training and testing data that allow engineers to compare their machine learning methods against each other. They are measurement devices that provide an estimate of how well AI software will perform in a real-world setting. Most are circulated publicly, while others are proprietary. The AI software that steered the car that killed Elaine Herzberg was most likely evaluated on several internal benchmark datasets; Uber has named and published information on at least one. More broadly, benchmarks guide the course of AI development.

They are used to establish the dominance of one approach over another, and ultimately influence which methods get utilized in industry settings.

The single most important benchmark in the field of computer vision, and perhaps AI as a whole, is ImageNet. Created in the late 2000s, ImageNet contains millions of pictures—of people, animals, and everyday objects—scraped from the web. The dataset was developed for a particular computer vision task known as "object recognition." Given an image, the AI should tag it with labels, such as "cat" or "dog," describing what it depicts.

"Datasets have hidden and complicated histories."

It is hard to overstate the impact that ImageNet has had on AI. ImageNet has inaugurated an entirely new era in AI, centered on the collection and processing of large quantities of data. It has also elevated the benchmark to a position of great influence. Benchmarks have become *the* way to evaluate the performance of an AI system, as well as the dominant mode of tracking progress in the field more generally. Those who have developed the best-performing methods on the ImageNet benchmark in particular have gone on to occupy prestigious positions in industry and academia. Meanwhile, the AI systems built atop of ImageNet are being used for purposes as varied as refugee settlement mapping and the identification of military targets—including the technology that powers Project Maven, the Pentagon's algorithmic warfare initiative.

The assumption that lies at the root of ImageNet's power is that benchmarks provide a reliable, objective metric of performance.

This assumption is widely held within the industry: startup founders have described ImageNet as the "de-facto image dataset for new algorithms," and most major machine learning software packages offer convenient methods for evaluating models against it. As the death of Elaine Herzberg makes clear, however, benchmarks can be misleading. Moreover, they can also be encoded with certain assumptions that cause AI systems to inflict serious harms and reinforce inequalities of race, gender, and class. Failures of facial recognition have led to the wrongful arrest of Black men in at least two separate instances, facial verification checks have locked out transgender Uber drivers, and decision-making systems used in the public sector have created a "digital poorhouse" for welfare recipients.

Benchmarks are not neutral pieces of technology or simple measurement devices. Rather, they and the measures that accompany them are situated, constructed, and highly value-laden—the reality of which is frequently discounted or ignored in dominant AI narratives. Datasets have hidden and complicated histories. Uncovering these histories, and understanding the various choices and contingencies that shaped them, can help illuminate not only the very partial and particular ways that AI systems work, but also help identify the upstream origins of the harms they produce. What we need, in other words, is a genealogy of benchmark datasets.

Chair Inherits from Seat

Teaching computers how to see was supposed to be easy. In 1966, the AI researcher Seymour Papert proposed a "summer project" for MIT undergraduates to "solve" computer vision. Needless to say, they didn't succeed. By the time the computer scientist Fei-Fei Li entered the field in the early 2000s, researchers had acquired a much deeper appreciation for the complexity of computer vision problems. Yet progress remained slow. In the intervening decades, the basics had been worked out. But there was still far too much manual labor involved.

At a high level, computer vision algorithms work by scanning an image, piece by piece, using a collection of pattern recognition modules. Each module is designed to recognize the presence or absence of a different pattern. Revisiting our cat-detector model, some of the modules might be sensitive to sharp edges or corners and might "light up" when coming across the pointy ears of a cat. Others might be sensitive to soft, round edges, and so might light up when coming across the floppy ears of a dog. These modules are then combined to provide an overall assessment of what is in the image. If enough pointy ear modules have lit up, the system will predict the presence of a cat.

When Li began working on computer vision, most of the pattern recognition modules had to be painstakingly handcrafted by individual researchers. For computer vision to be effective at scale, it would need to become more automated. Fortunately, three new developments had emerged by the mid-2000s that would make it possible for Li to find a way forward: a database called WordNet; the ability to perform image searches on the web; and the existence of crowdworking platforms. Li joined the Princeton computer science faculty in 2007. There, she encountered Christiane Fellbaum, a linguist working in the psychology department, who introduced her to a database called WordNet. WordNet, developed by cognitive psychologist George A. Miller in the 1980s, organizes all English adjectives, nouns, verbs and adverbs into a set of "cognitive synonyms ... each expressing a different concept." Think of a dictionary, where words are assembled into a hierarchical, tree-like structure instead of alphabetically. "Chair" inherits from "seat," which inherits from "furniture," all the way up to "physical object," and then to the root of all nouns, "entity."

Fellbaum told Li that her team wanted to develop a visual analog to WordNet, where a single image was assigned to each of the words in the database, but had failed due to the scale of the task. The resulting dataset was to be called ImageNet. Inspired by the effort, in early 2007 Li took on the name of the project and made it her own. Senior faculty at Princeton discouraged her from doing so. The task would be too ambitious for a junior professor, they said. When she applied for federal funding to help finance the undertaking, her proposals were rejected, with commenters saying that the project's only redeeming feature was that she was a woman.

Nonetheless, Li forged ahead, convinced that ImageNet would change the world of computer vision research. She and her students began gathering images based on WordNet queries entered into multiple search engines. They also grabbed pictures from personal photo sharing sites like Flickr. Li would later describe how she wrote scripts to automate the collection process, using dynamic IP tricks to get around the anti-scraping safeguards put in place by various sites. Eventually, they had compiled a large number of images for each noun in WordNet.

However, they still needed a way to verify that the images actually matched the word associated with them—that all of the images linked to "cat" really showed cats. Since the scraping was automated, manual review was required. This is where Amazon's crowdworking platform, Mechanical Turk (MTurk), came in. It was a "godsend," Li later recalled. MTurk had been launched just a couple of years before, in 2005. Her team used it to hire workers from around the world to manually review the millions of images for each WordNet noun and then verify the presence or absence of a target concept.

The ImageNet dataset would take two and a half years to build, its first version completed in 2009. When it was finished, it consisted of fourteen million images labeled with twenty thousand categories from WordNet, including everything from red foxes to Pembroke corgis, speed boats to spatulas, baseball players to scuba divers. At the time, it was the largest publicly available computer vision dataset, hosted on the ImageNet website for anyone to download.

Convolutional Cat Ears

Although it took an immense amount of effort to create ImageNet, the initial uptake was slow. Li and her students presented a poster announcing its creation at a major computer vision conference. Tucked away in a corner of a conference center in Miami Beach, they even distributed logoed keychains and pens to advertise it. But beyond ImageNet's limited popularity, there was a deeper issue. The problem that Li had hoped to solve with the creation of ImageNet—the fact that object recognition modules needed so much manual work to produce—still hadn't been solved.

In an attempt to encourage wider adoption of the dataset, Li's team decided to organize a competition. The ImageNet Large Scale Visual Recognition Challenge was officially launched in 2010. To enter the challenge, competitors would develop machine learning models using the benchmark training data, and submit their model's predictions on a set of the benchmark testing data. The team whose model could detect objects in the images with the highest accuracy would be the winner.

In 2012, computer scientist Alex Krizhevsky, along with his colleagues at the University of Toronto, won the ImageNet Challenge with AlexNet, a neural network–driven machine learning model that outperformed all other competitors by a previously unimaginable margin. After a long period during which neural networks were out of fashion in AI, AlexNet almost single-handedly put them back at the center of research into the field. Part of what enabled the return of neural networks was the much greater processing power of modern computers, which was needed to handle massive datasets.

"Although it took an immense amount of effort to create ImageNet, the initial uptake was slow."

AlexNet helped fulfill the potential of ImageNet and solve the problem that Li had identified when she first started out: that computer vision required too much manual labor. Krizhevsky and his colleagues didn't rely on handcrafted modules for object recognition. Rather, using neural networks, AlexNet was able to "learn" what an object looked like completely from the data.

Neural networks work by stacking layers of artificial "neurons" on top of each other. Each layer alters the image slightly, like a camera lens filter. Some of the first layers of AlexNet's neural network model, known as "convolutional," allowed it to automatically encode information that used to be manually coded—like the pointy edges of a cat's ears. There was no longer any need to enter such information by hand. With enough images of cats, the neural network would be able to learn which patterns were most predictive of the animal.

AlexNet's success is often credited with sparking the resurgence of neural networks—under the new name of *deep learning*, which refers to multiple stacks of neural network layers—as the dominant machine learning paradigm. The 2012 paper associated with the model now has over seventy-two thousand citations on Google Scholar, an indication of its popularity in academic and industry circles alike. Deep learning techniques have achieved near-universal adoption not only within computer vision, but also within natural language processing—which works with human language—and a number of related subfields.

The deep learning era has, in turn, placed data—more specifically, vast quantities of data—at the center of AI development. Because deep learning models become more accurate when trained on more data, tech companies are highly incentivized to gather as much data as possible. The amount of information available on the internet continues to grow. Users on Instagram share 8.9 million images a day alone. Meanwhile, a new cottage industry of data annotation work has sprung up to feed soaring demand for data labeling. The people who do this work are typically subcontractors or crowdworkers, like the MTurkers who helped create ImageNet, and represent a growing underclass of invisible tech workers.

Algorithms of Oppression

Why does the history of ImageNet matter? ImageNet has had an enormous influence on the field of modern AI, and on many of the AI systems that affect so many aspects of our lives. By understanding the particular circumstances of ImageNet's creation, we can better understand these systems. We can also understand how the progress of AI moves in fits and starts, how its reliance on massive amounts of data is contingent and accidental, and how its present course was just one possible path among many.

ImageNet was built on three technological pillars: WordNet, search engines, and crowdworking. The reliance on WordNet has proven to be particularly problematic. ImageNet recodes outmoded and prejudiced assumptions—many of them racist, sexist, homophobic, and transphobic—because those assumptions come directly from WordNet.

"Because deep learning models become more accurate when trained on more data, tech companies are highly incentivized to gather as much data as possible."

A good illustration of this comes from a website called ImageNet Roulette. Developed by AI researcher Kate Crawford and artist Trevor Paglen, ImageNet Roulette allows users to upload images of themselves. These images are then analyzed by a machine learning model, trained on a set of ImageNet data, which generates a description. When *Guardian* journalist Julia Carrie Wong uploaded a photo of herself, it labeled her with an ethnic slur, while *New York Times* video editor Jamal Jordan was consistently labeled as "Black, Black person, blackamoor, Negro, or Negroid," no matter which image he uploaded.

To their credit, ImageNet's creators quickly sanitized the dataset of such labels for its future users. But those categories still exist in multiple machine learning systems, due in part to the influence of ImageNet. AI researchers Vinay Uday Prabhu and Abeba Birhane recently demonstrated that the categories in the WordNet database persist in several widely cited public computer vision benchmarks, resulting in the takedown of a prominent benchmark by MIT called Tiny Images. And if they exist in these open datasets, then they are potentially replicated in many internal industry ones.

WordNet is not the only issue with ImageNet, however. The data contained within ImageNet was gathered from internet search engines in the early 2000s. Such search engines, as the UCLA professor of information studies Safiya Umoja Noble has explained, encode racist and sexualized imagery for Black, Latina, and Asian women, and overrepresent imagery of white men in positions of power. These engines also portray a Western white male vision of the world, associating "beauty" with white women, "professor" or "ceo" with white men, and "unprofessional hairstyles" with Blackness. These assumptions filtered into ImageNet as the dataset was constructed.

One common response from AI researchers to the oppressive aspects of ImageNet, and to the crisis of algorithmic injustice more generally, is that the problem lies with the data: if we get more or different data, then all these problems will inevitably go away. This was the response that Yann LeCun, one of the "godfathers" of deep learning and chief AI scientist at Facebook, provided when a machine learning model designed to depixelate faces ended up whitening them as well. Timnit Gebru, co-lead of Google's Ethical AI team, struck back, underscoring how AI systems cause real harm and exacerbate racial inequality, and arguing that improving them must mean more than just focusing on better data collection. (Disclosure: two of us, Hanna and

Denton, are members of Gebru's team.) *[Eds.: Shortly before this issue went to press, Gebru was terminated by Google.]* Furthermore, data collection efforts aimed at increasing the representation of marginalized groups within training data are often executed through exploitative or extractive mechanisms such as, for example, IBM's attempt to "diversify" faces by scraping millions of images from Flickr without the consent of people in them. As Gebru explained during a tutorial at the Computer Vision and Pattern Recognition conference in June 2020, "Fairness is not just about datasets, and it's not just about math. Fairness is about society as well, and as engineers, as scientists, we can't really shy away from that fact."

> *"One common response from AI researchers to the oppressive aspects of ImageNet, and to the crisis of algorithmic injustice more generally, is that the problem lies with the data: if we get more or different data, then all these problems will inevitably go away."*

A particularly pernicious consequence of focusing solely on data is that discussions of the "fairness" of AI systems become merely about having sufficient data. When failures are attributed to the underrepresentation of a marginalized population within a dataset, solutions are subsumed to a logic of accumulation; the underlying presumption being that larger and more diverse datasets will eventually morph into (mythical) unbiased datasets. According to this view, firms that already sit on massive caches of data and computing power—large tech companies and AI-centric startups—are the only ones that can make models more "fair."

> *"We need to develop genealogies of data to show that datasets are the product of myriad contingent assumptions, choices, and decisions and could in fact be otherwise."*

A Genealogy for the Many

Exploring the history of ImageNet has implications not only for how we discuss the problems and failures of AI, but also for how we make critiques and formulate solutions to those issues. We need to develop genealogies of data to show that datasets are the product of myriad contingent assumptions, choices and decisions, and that could, in fact, be otherwise. Genealogy is an interpretive method of analysis, which we can apply to the historical conditions of dataset creation. Understanding these conditions illuminates the origins of certain problems, but it also opens up new paths of contestation by enabling us to imagine new standards, new methods for evaluating AI progress, and new approaches for developing ethical data practices in AI.

Instead of the narrow focus on "bias," we can start to ask deeper questions such as: How did particular datasets emerge? What agendas, values, decisions, and choices governed their production? Who collected the data and with what purpose? Are the people represented in the datasets aware that they are participants in them? Can they meaningfully opt out? How about the workers, like the Amazon MTurkers, who annotated them? Were they fully recognized for their labor and fairly remunerated? And, most importantly, does the creation of the datasets serve the interests of the many or only those of the few? ∿

Alex Hanna, Emily Denton, Andy Smart, and Hilary Nicole are researchers at Google. Razvan Amironesei is a research fellow at the University of San Francisco.

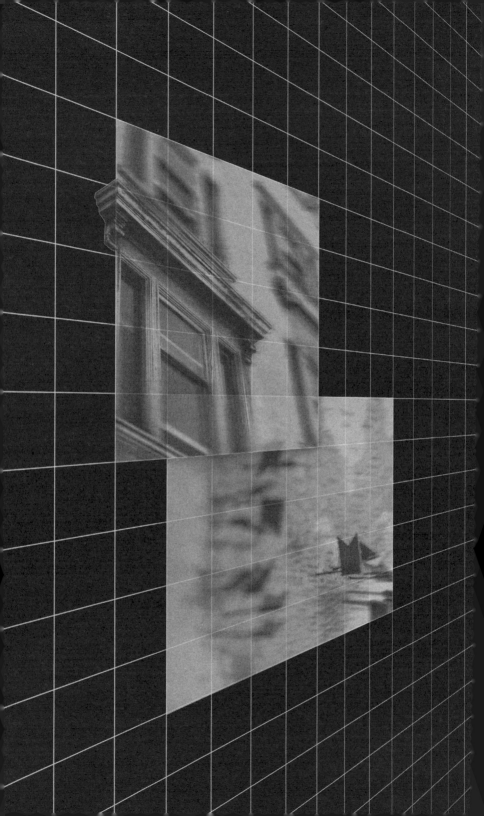

Evictor Structures

Erin McElroy and Azad Amir-Ghassemi on Fighting Displacement

The Anti-Eviction Mapping Project (AEMP) describes itself as "a data-visualization, data analysis, and storytelling collective documenting dispossession and resistance upon gentrifying landscapes." That compact summary emerges out of projects that sprawl across seven years and now three chapters—the group's original San Francisco Bay Area chapter, as well as newer chapters in New York and Los Angeles. Volunteers in each city have collaborated with tenants unions and housing justice groups in those places to produce dozens of oral histories, maps, reports, and zines.

Before there were chapters in other cities, AEMP was only a Bay Area project, and members of the collective met in the San Francisco Tenants Union building. When Erin McElroy started the project in 2013, San Francisco was ground zero for the second tech boom that saw the tech industry rapidly reorganize the city and displace longtime residents. That year, Rebecca Solnit published an essay comparing Google buses to the invading Prussian army during the siege of Paris. McElroy was a housing organizer thinking about how mapping evictions might be a way to wrest some power from the disruptors and redistribute it to tenants.

Today, AEMP members from all three chapters are building EvictorBook, the website we might have gotten instead of Facebook fifteen years ago if Mark Zuckerberg had been more interested in housing justice than in rating women's faces. The site lets you search for evictors by name or building address and surfaces eviction data that's almost impossible for tenants to get themselves. We sat down with McElroy and AEMP-LA member Azad Amir-Ghassemi to learn more about the project.

––––––

Why did you make EvictorBook?

Erin McElroy: Particularly since 2008, we've seen the rise of corporate landlordism. We've seen huge investment companies—Blackstone and Invitation Homes are two of the biggest in the US, but there are many others—that will buy up swaths of property with unique limited liability company (LLC) or limited partnership names.

Take 55 Dolores Street LLC and 49 Guerrero Street LLC. These two LLCs are subsidiaries of Urban Green Investments, which is a big investment company that evicted many tenants in San Francisco in 2013. In the process of buying those properties, the company established a separate LLC for each one, which helped them with a number of things in terms of finance and liability, and also afforded them anonymity. It's often very hard for tenants to know which other buildings in the city are owned by their landlord if each property has a unique-sounding ownership name.

When that was happening in 2013, the Anti-Eviction Mapping Project was just getting started, and we didn't have a tool like EvictorBook, so we were doing property research manually. We'd create static websites where we'd list all the LLCs and all the evictions that we were able to connect to that investment company, with the idea that this information should be public and that tenant organizers should be able to use it for campaigns—ideally, multibuilding campaigns against large-scale landlords. They were essentially profile pages on different landlords. That's useful because you have a much stronger chance of winning a fight against your landlord if you're working together with other tenants across that landlord's other buildings.

We also created a lookup tool and pledge map back in 2014. You can type in an address, see if there's been an eviction there, and then pledge to not rent from that landlord. The map was pulling in public eviction data from the San Francisco Rent Board. The obvious next step was to connect other data to it, like parcel data and corporate ownership data. That turned out to be a lot harder than we realized, so it's taken us some time. But that's what the EvictorBook website does: it brings together a lot of those tools that we've already been using for years and makes it easier to see the landlords, LLC networks, and eviction histories of different buildings in San Francisco. You can enter an address or a landlord's name, and we'll show you a profile page with evictions connected to that entity.

How did you all build it?

Erin: The work began in early 2019 in collaboration with other member organizations of the San Francisco Anti-Displacement Coalition (SFADC), and in collaboration with the Mapping Action Collective (MAC) in Portland, Oregon.

Since then, we've done some workshops with different tenant groups to figure out what needs and questions they had given the new Covid conditions, and we've been really lucky to get new frontend people involved in development work. Right now, we're

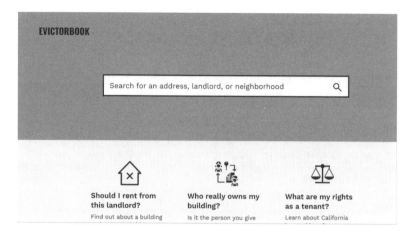

EvictorBook can be searched by entering an
address, landlord, or neighborhood name.

making the site more user-friendly and doing more user testing
in LA and Oakland.

Azad Amir-Ghassemi: The tenants unions in Los Angeles and
San Francisco already do this kind of research when they're
fighting evictions. And organizers in other cities around the
country are making similar tools, drawing from the same
resources that we draw from: assessor data, property records,
eviction data where possible, and sales data.

> *"In the process of connecting all
> the data we're working with, we're
> also trying to map networks of
> financial speculators and evictors."*

With EvictorBook, we're just trying to make it easier, since tenants are most motivated to do this research when they're already in crisis. They're faced with belligerent landlords who harass them in all kinds of ways—ripping out appliances or sending fake legal notices to get people out.

Before we had EvictorBook, we'd start at the assessor's office doing searches on paper and then look up what we could online. But property ownership networks have gotten very sophisticated. People wring out profits from one city and then move to other cities. So in the process of connecting all the data we're working with, we're also trying to map networks of financial speculators and evictors.

Have you heard about any surprising uses of EvictorBook?

Azad: There are a bunch of use cases. One is proactively asking: What has this landlord done in the past? Is this landlord representing themselves accurately?

Another is: Is this landlord lying to me about what's taking place? In California, landlords can do what's called an Ellis Act eviction, where they evict someone because they say they no longer want to rent out the building. In Los Angeles, there's a five-year limit with this kind of eviction when the landlord is not allowed to rent out the unit they evicted someone from. But they can move in for a little while and then sell the property before five years are up. There are also owner move-in evictions that operate similarly. EvictorBook could tell you if something like that is the reason you're being evicted.

Do landlords do these kinds of evictions in order to get rid of rent control in their buildings?

Erin: Indirectly, yes, because they can rent and sell the units for more money if there's no rent control. In San Francisco, after a landlord uses the Ellis Act to evict tenants from a rent-controlled building, the building will often then get sold as multiple "tenancy in commons," which will still have rent control—but then *they* will get converted into condos. And when that happens, the building loses rent control. That's one of many loopholes. In short, what we're seeing with Ellis Act evictions and Owner Move-In evictions is that we're losing effective rent control, which, in the case of condo conversion, is a nonrenewable type of protection.

Azad: I was thinking we could add a feature to EvictorBook that shows how much a landlord profited from an eviction. We have the sale price before and the sale price after.

Erin: Oh, yeah, that would be great to see.

Azad: It wouldn't be hard to do. We have all this data that we're actually not spending much time analyzing. We know who the

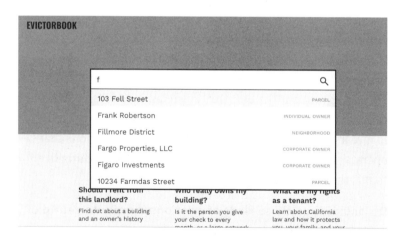

For multibuilding campaigns, tenants have to know what other properties their landlord owns. EvictorBook surfaces that information.

evictor of a given building was — and not just who evicted that one unit a few years ago, but looking back over fifteen, twenty years of evictions. That, combined with the networks of LLCs, shows not just ownership structures but also the *evictor structures* in the history of housing, which is the history of gentrification and the history of displacement in neighborhoods that are financialized. There are so many questions we could use this data to think about.

Half-Open Data

Erin: There are all these public datasets and open data portals that different cities have, but none of those datasets will indicate who corporate landlords or evictors are.

Azad: Or the stories of the people who are impacted.

Erin: Right. So the data that's fueling EvictorBook is technically public, but people have to put zillions of hours into stitching it together for what we need. You would think that cities and different administrative bodies would do this work, but it's not in their interest. So we're taking public data and connecting it so that it can be useful to tenants.

I remember looking for rent stabilization data when I lived in New York, and the closest I found was this project, AmIRentStabilized.com, which is someone's personal project to help people do this. The city has that data and could share it with tenants, but they don't.

Erin: Totally.

Azad: And New York is a special case where there are different levels of rent stabilization and rent control, and the property owners know the status of their building, but it's not easy for tenants to look up that information themselves.

The website walks you through the process of contacting a city office, which is then supposed to mail you the answer, but if they don't, you're supposed to set a calendar reminder for yourself to follow up with them!

Azad: That's because these institutions are beholden to property owners who fight tooth and nail for data about them to remain inaccessible. And with open data, as soon as the data gets politicized, it gets pulled. We just saw this in Chicago, where there were some open datasets about policing in the city—several reports were written about them—and then that open data got removed. The neoliberal fantasy that motivated the open data movement in the first place was like, "If you just open the data, people will make great business products out of it!" But it's proven to be much more complicated.

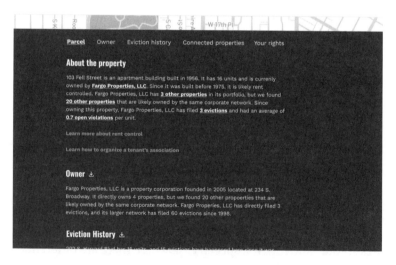

EvictorBook shows historical eviction data for a given building — data that's otherwise difficult to find and piece together.

It's also a ton of ongoing work to maintain that data and keep producing clean datasets. We want to be very transparent in EvictorBook. We're not just saying: this person might own this property. We're saying: there was an eviction here, this person owned the property during that eviction, and here's why we're sure of that. A ton of due diligence goes into that.

We've talked a lot over the course of this project about how we can avoid replicating some of the horrible data practices of property tech. Landlords have blacklists on prospective renters that they create by doing a regex on five characters of a last name, without any other verifying information. [*Eds: Regex is short for regular expression, a way of defining a pattern and then finding matches that fit the pattern.*] That's a great way to block large swaths of entire demographic populations from renting. We're being extremely careful about that kind of thing.

> *"With open data, as soon as the data gets politicized, it gets pulled."*

Tell me more about these lists of renters based on five characters of a last name. I usually think of landlord tech more as surveillance cameras in buildings.

Azad: It's that too.

Erin: Tenant screening has been around since the mid-1970s and has experienced different booms, such as a big one after 9/11 and another one after 2008. It's getting more and more robust as an industry, weaving together past eviction history with any criminal record history that someone might have, along with credit reporting. There are so many bad data practices within each step. Of course, people are being wrongly criminalized at the outset, but also, oftentimes, these screening bureaus only look to see if a tenant has *appeared* in any kind of housing court—not necessarily what the outcome of the case was. So someone may have reported that their landlord was abusive, which gets their name registered in some database because they filed a complaint. It's a mess.

Azad: We talked to eviction lawyers who were shocked that those lists were out there. We assumed that they knew, but they were like, "No, there is no central repository." There's just a patchwork of data on people who have been evicted. I think tenant screening companies literally just sell CSVs back and forth.

Erin: Different states have different policies about this. A few years ago, California passed a law that's supposed to protect tenants from having their data collected by these screening companies. New York just last year passed a law that bans tenant blacklists. But apparently it's still happening in both California and New York despite these laws. More broadly, there's not a lot of regulation or enforcement around how these companies get data. In New York, LexisNexis sends its own people to housing court to take photos of the computers with all the eviction records. Then they bring those back, standardize them, and sell the data to tenant screening companies. Really weird things like that!

The surveillance you mentioned is also part of the property technology, or *proptech*, umbrella, which involves biometric data and all kinds of other tools and practices. We have been having a lot of conversations with other organizers about how to ensure that we don't put data out there that can be used against tenants.

That gets to a security issue: there are housing justice groups already using EvictorBook, but it's not publicly available yet. Can you talk about how you're thinking about whether or not to make it public?

Azad: We've been thinking through worst-case scenarios: how could this data be used in a way that we didn't plan for? What if the data was hacked? The eviction data is something we have access to that most people do not. We've gotten it through relationships with the courts, housing boards, and tenant organizers. So the raw data itself will probably remain restricted. But the rest of the site will be public and usable.

Erin: We've been weighing the risks of putting it out there—there's always more verification of the data that can be done. On the other hand, there are risks involved in *not* putting it out there—it could be useful to people organizing rent strikes right now, particularly with Covid.

Anti-eviction Constellation

Chapters are a relatively new thing for AEMP. AEMP used to be limited to San Francisco. As the group grows, how do you keep the focus on housing justice that the original chapter has?

Erin: One of the things that I've always valued about AEMP is that people come in with such different backgrounds. Probably half of the people in the collective are doing narrative-based and storytelling work. There is a growing crew that's bringing software skills, which are necessary—but we do work collectively to stay as far away as possible from Silicon Valley tech culture in our processes, in how we conceive our projects, and in how we work with community partners. The housing justice politics undergirding our work are first and foremost. Many people in the project are also in tenants unions, show up in different coalition meetings, and/or participate in actions.

We're also a young organization. When we began in 2013, our group was housed in the San Francisco Tenants Union, which has been around for decades, as have many other orgs we've worked with. When the LA and New York chapters started, people in those places did a lot of outreach to figure out what kind of organizing was already happening in their cities.

Azad: In every chapter, there's pretty tight integration with tenants unions. I know in LA there are four or five chapters across the city, and we have people going to meetings at most of those. We see ourselves as beholden to that movement; it's built into our projects, the community agreements we've written, and the onboarding. We're very clear that we're not just building whatever tech.

The Anti-Eviction Mapping Project is always juggling collaborations with multiple other groups. At the same time, both of you have mentioned the zillions of hours that have gone into EvictorBook alone. How are you sustaining these efforts in a volunteer project?

Azad: I don't know how Erin keeps it all together. I'm on two projects and it's a full-time commitment.

Erin: You do so much, Azad. He's working on both EvictorBook and the COVID-19 Global Housing Protection Legislation and Housing Justice Action Map, which tracks tenant protections, housing justice actions, landlord retaliation, and soon will include tenants' oral histories since Covid.

It's an interesting moment because, for a while, we had these three chapters, and each chapter was working on its own projects, so I don't even know a lot of what's happening in places where I'm not there, going to those meetings. I know in the Bay, there's a thirty-minute film that's an homage to tenant organizing that weaves together three struggles: the anti-Google fight in San Jose, the fight for rent control in Santa Rosa, and the fight for Aunti Frances's home—she's a former Black Panther who was evicted from her home in Oakland a few years ago. Our *Counterpoints* atlas will come out later this year with PM Press. That's a project that many, many people have been working on for almost three years. Probably well over a hundred people contributed to the atlas, some affiliated with AEMP and others not. There's also the *Black Exodus* zine that came out last year that tracks Black experiences of gentrification and resistance in San Francisco. I'm sure Azad knows about a lot of projects in LA that I don't know about.

> *"We see ourselves as beholden to the housing movement. We're very clear that we're not just building whatever tech."*

With EvictorBook, we started this interchapter collaboration model that more of our work is taking on, since Covid means that we're all meeting online and not in person anyway. We're having interchapter meetings every month whereas we used to have those every few months. But, yeah, it's all volunteers, and we're always trying to find ways to create more sustainability and be better organized. People are dedicated, and it somehow keeps going.

Azad: I think it's partly that these are decentralized projects, and we're just providing a framework. The framework is: How does this relate to housing justice? You're not just making this for your own curiosity or to scratch your own itch; you can do that by yourself. So what is the nexus between how we see our role in the housing movement and the kinds of group projects that people are willing to take on?

When the LA chapter started, we would pair with another organization in the movement and they would say, "Help us look into this." More recently, it's become less of that model. Instead, we have a north star—it might be a little vague—and within that project, people are decentralized and empowered to work toward the ends they define. That's the post-and-beam structure that AEMP provides.

What would you say is the north star of all of these projects? What do you see them as building toward?

Erin: This is not an AEMP-affiliated project in any way, but we're really inspired by the Moms for Housing movement, which has shifted the conversation in the Bay Area. Four Black mothers in Oakland who had been unhoused organized to reclaim a home that was sitting vacant. It was owned by Wedgewood, which is one of these big investment companies with many LLCs throughout California. They were asking: Why should there be this vacant home owned by this corporate landlord when there are so many people who need housing now? The mothers helped create a huge movement in Oakland that inspired people in LA and beyond to do similar organizing. That spirit, that politic, has all sorts of afterlives, particularly in this intense moment where eviction moratoriums are ending while there's still a pandemic going on.

Azad: What's the north star? That's a question that I think everyone who does work like this asks themselves. Is it about resisting, is it about building toward something new, is it about throwing sand in the gears? For me, it's a mix of all of the above: community land trusts, decommodified housing—and let's also repeal the state proposition that bans new public housing from being built in California! But more broadly, I think the idea is to keep making constellations. We're providing north stars for others, and they provide them for us. 〰️

Under a Blood-Red Flag

by Nayantara Ranganathan

A new emoji furthers Hindu supremacism—and Silicon Valley
is to blame.

———————

In March 2019, the Unicode Consortium, which controls the publication of emojis worldwide, released an emoji of a Hindu temple. Until then, a Hindu temple had been conspicuously missing
from the set of emojis representing religious places, which
included a Christian church ⛪ and Shinto shrine ⛩ (approved
as part of Unicode 5.2, in 2009) and a kaaba 🕋, mosque 🕌, and
synagogue 🕍 (Unicode 8.0, 2015). The most popular emojis are
the ones that convey emotion, such as the crying-while-laugh-
ing face 😂, but many emojis are also powerful ways to represent
cultures and identities (a woman in a headscarf 🧕, same-sex
couples 👬, different skin tones 👋👋👋👋👋). On the face of it,
then, the publication of an emoji evoking the religion of 1.2
billion people seemed like an important act of inclusion.

But the form that the temple emoji took, and the timing of its
release, carried distressing political connotations. Its publication coincided with a period of intensifying demands for a Hindu

temple to be built on a contested site in the northern Indian city of Ayodhya. The construction of this temple is one of the most fraught political projects in contemporary India, and a symbol of violent attempts by Hindu supremacists to create a Hindu-first nation at odds with the ideal of a secular Indian democracy that emerged after Indian independence, in 1947.

In 1992, Hindu fundamentalists from all over India traveled to Ayodhya to demolish a sixteenth-century mosque called the Babri Masjid. In the following months, at least two thousand people, mainly Muslims, were killed in violence between Hindus and Muslims across India. Much of this violence was led or inspired by the Rashtriya Swayamsevak Sangh (RSS), a militant Hindu nationalist organization that boasts millions of members throughout the country. The RSS has a two-pronged mission: to narrow the extraordinarily broad set of beliefs and practices known as Hinduism into a single political form of the religion, and to make India—which is roughly 80 percent Hindu, 15 percent Muslim, and 5 percent Christian, Sikh, Buddhist, Jain, and others—a formally Hindu state.

"The emoji bears two emblems of Hindu fundamentalism: the color saffron and a distinctive red flag."

If the destruction of the Babri Masjid was one of the RSS's greatest symbolic victories, its greatest political victory came in 2014, when Narendra Modi, a longtime officer of the group, was elected India's prime minister. Since then, the country has taken a brazenly fundamentalist turn, with mob killings of Muslims being implicitly and explicitly encouraged by members of Modi's political party, the Bharatiya Janata Party (BJP), which is effectively the political wing of the RSS.

The mission to build a Hindu temple in Ayodhya has also advanced. In 2017, an extremist member of the BJP took power in India's largest state, Uttar Pradesh, where the contested religious site is located. He promised that under his leadership, no force could stop the construction of the temple. Then, in November 2019, the long-awaited verdict in a twenty-seven-year-old court case was decided by India's highly politicized supreme court: the construction of the Hindu temple, known as the Ram Mandir, could go ahead.

This was the charged political climate into which the Hindu temple emoji was released by the Unicode Consortium earlier that year. Remarkably, the emoji bears two of the main emblems of Hindu political-religious fundamentalism: the color saffron, and a distinctive red flag that looks exactly like the banner of the RSS. The emoji's design is also similar to the plans for the Hindu temple that is to be built in Ayodhya, which have been in development since the early 1990s.

The Hindu temple emoji can now adorn almost any message on any social media site in the world, whether or not users understand its significance. Many people, in India and abroad, put the emoji in their screen names to signal their allegiance to a Hindu-first nation; some also use it when making calls for violence against Muslims. But even when the emoji is used in less explicitly political contexts, its effect is to uphold and normalize the RSS's political version of Hinduism, its violent attacks on Muslims, and its Hindu-supremacist vision of India.

Because it has a fairly superficial theory of what emojis are and remains focused on a narrow set of criteria for approving new emojis, the Unicode Consortium failed to understand the cultural and political significance of the Hindu temple emoji. The story of the emoji's development reveals that the process for approving new emojis—arguably the most popular lingua franca in history—privileges the economic concerns of large tech companies, and ultimately replicates the ways that these companies see, and fail to see, the world.

Hello, Interoperator?

When you type a message into your smartphone or scroll through the articles on your Facebook feed, you are interacting with the work of the Unicode Consortium. The consortium is responsible for encoding all the characters you see on your screen—letters and numbers in various scripts, hanzi and kanji, dingbats—into binary, so that they can be read on pretty much any machine, with any operating system, anywhere in the world. The purpose is to ensure that a line of code written in Bangalore, or a tweet fired off from San Francisco, arrives at its destinations essentially unchanged. Want a functioning Bengali website that can be accessed just as easily in East London as in Dhaka, or a Chinese-manufactured tablet that can render fonts designed in Accra? You need Unicode for that.

> *"By encoding emojis, the Unicode Consortium exerts a powerful form of sovereignty over digital life."*

By encoding many of the world's scripts—Latin, Arabic, Greek, Tamil—in this way, the Unicode Consortium plays a crucial role in determining who can use the internet, which languages will survive digitization, and who can reap the gains of the digital age. It also helps to determine who can enter the global digital marketplace as consumers, advertising targets, and data sources for extractive surveillance capitalism. In other words, universal standards for interoperability are not just about bringing people online and connecting them—they're also about driving profits through an ever-expanding digital ecosystem.

These goals reflect the structure of the consortium. The consortium is a Silicon Valley–based nonprofit, and many of the people who do its work—assessing new alphabets for inclusion, doing the codification—are volunteers. But the power to decide what gets encoded by the consortium ultimately rests with the top tier of its paying membership, a group of about eight of the world's largest tech companies (including Apple, Facebook, Google, and Microsoft) along with a small handful of governments and a university, which buy into the consortium for roughly $10,000 to $20,000 per year, and get voting power as a result. (By paying as little as $35, other individuals and institutions can be members of the consortium, but they don't get voting privileges.)

Since 2007, as part of its overarching aim to expand the use of digital technologies, the consortium has taken on the standardization of emojis. Emojis not only help encourage user engagement with consumer technologies, but also provide a valuable seam of minable data—for example, in the millions of emoji reactions to Facebook posts that occur every day. As of Unicode 13.1, released in September 2020, there are 3,521 emojis, as well as more than 143,000 characters in dozens of different scripts. By encoding new emojis, the consortium is deciding which characters can exist in this new visual language. This gives Unicode a powerful form of sovereignty over digital life, and adds a further political dimension to the consortium's work, since it decides what sorts of representation — interracial couples, say — get universalized as emojis, and which do not.

The consortium itself is not always clear on or honest about the significance of its work. It tends to see emojis as "playful, colourful representations," as one Unicode document puts it. It also likes to present the process of creating new emojis as a fairly open one. It's true that, by submitting a formal proposal, including a provisional design, anyone can suggest a new emoji to the organization, out of which about sixty to seventy new emojis are approved every year. However, the proposals are evaluated behind closed doors by the consortium's Emoji Subcommittee, and then voted on by the highest tier of corporate, state, and institutional members. Although there are criteria that new emojis supposedly must fulfill, the extent to which an emoji does meet those criteria is a matter of broad interpretation. What's more, the criteria are designed to ensure that an emoji is as widely used as possible, so that tech companies can derive the maximum monetary value from it. A narrow focus on these standards can obscure the larger cultural and political significance a new emoji might bear.

To maintain its unique role in encoding the world's language scripts, the consortium presents itself as neutral and above politics. It tries to avoid political conflicts by admonishing designers not to "justify the addition of emoji because they further a 'cause,' no matter how worthwhile." All the same, "a proposal may be advanced *despite* a 'cause' argument — if other factors are compelling," states the document detailing how to submit emoji proposals, which was written in 2009 by an Apple employee. As the scholars Luke Stark and Kate Crawford observed in a 2015 paper, "proposed solution[s] for improving emoji diversity in fact [signal] a further evolution in the business models of affective digital communications." In other words, the political and cultural representation furnished by emojis — brown skin tones, a trans rights flag ▬ — may be deeply meaningful to users, but for the Unicode Consortium and its members, allowing such representation is primarily a means to get more people to use more digital technology more often.

Bright Orange to Blood Red

The Unicode notation for the Hindu temple emoji is U+1F6D5. In the Twitter version of the emoji, the graphic has a stepped structure, tapering upward in tones of saffron to a tall spire, with a prominent two-fanged flag at the top.

Its architectural form is an ambiguous mix of the two most canonized styles of Indian temple architecture, the north Indian Nagara and south Indian Dravidian. In this way, it seems to combine two major streams of ancient temple architecture, each with many offshoots and tributaries, into a single statement. It also shares many features of the temple that is to be built on the contested site in Ayodhya, according to a design that has recently been circulated on social media by the Modi government's Shri Ram Janmbhoomi Teerth Kshetra, a trust constituted to lead the construction and management of the temple.

The emoji was originally designed by Girish Dalvi, a professor at the Industrial Design Centre of the Indian Institute of Technology, and Mayank Chaturvedi, who at the time was working for an agency of the Maharashtra state government that is dedicated to promoting Marathi language and culture, including through Unicode representation. A first version of the proposal was sent back by the Emoji Subcommittee for clarifications about whether the proposed image would mean "Hindu temple" to anyone who saw it, and whether it would be the best

Hindu temple emoji.

symbol to account for regional differences in what temples look like. In response, Dalvi and Chaturvedi argued that their design used a "common architectural grammar" among Hindu temples, and that the saffron color provided "semantic reinforcement" because the "saffron colour is often associated with Hinduism." They also claimed that, like their design, "most Hindu temples have a flag at the apex."

This rationale used Unicode's criteria to justify the inclusion of the emoji's most blatantly Hindu-supremacist symbols. In reality, there's an extraordinary diversity to Hindu temple architecture. Many temples are whitewashed, while others are brilliantly polychromatic. In recent years, though, the color saffron—in hues from bright orange to blood red—has become increasingly associated with Hindu nationalism. There's even a word, "saffronization," for the Hindu right's attempts to rewrite everything from daily news to the history of the country in a way that reinforces Hindu supremacism: rechristening cities and streets that had Islamic-sounding names, purging textbooks of accounts of Mughal rule and heritage, claiming that ancient Hindus had "airplanes, stem cell technology, and the internet," and generally presenting a glorified version of the past in which Hindu civilization is always virtuous and always supreme. Online, saffron is used in a range of ways by members of the Hindu right to signal their allegiance to the idea of a Hindu-first nation, and to issue threats to non-Hindus—for example, by saturating images with saffron tones, or by using saffron as the background color for aggressive text memes.

The red swallow-tailed pennant that flies from the top of the temple emoji is also quintessentially Hindu-supremacist. It isn't necessarily the case that most temples have flags, and the temples that do have flags rarely have a flag of this kind. The pennant in the emoji, however, looks exactly like the *bhagwa dhwaj*, the banner of the RSS. Although it doesn't fly over many temples, it is sometimes used by Hindu fundamentalists to distinguish Hindu shops and other buildings from those owned by non-Hindus — information that could be put to vicious use during a pogrom, and which has come to inspire fear in non-Hindus in daily life.

> *"The emoji is a small but potent example of the way that Silicon Valley has chosen to make money from hate."*

The Unicode Consortium's criteria for new emojis require that the designs should be at once unique and representative, so that, for example, the beer mug emoji ▮ is instantly recognizable as a mug of beer while at the same time standing for beer in general. Part of what this means is that there will likely never be another Hindu temple emoji that could exist alongside, and contest the symbolism of, this one. The major social media platforms, software companies, and hardware manufacturers all adapt Unicode's approved emoji designs to their own products. Some of their versions of the temple are more clearly Hindu-supremacist than others, but all share at least one feature in common with the emoji's original design: many have pennants, and most are saffron. The upshot is that the Unicode Consortium's decision has massive downstream effects: whenever anyone adds a temple emoji to their Facebook post, they're reproducing a symbol of Hindu supremacism. (Members of the Unicode Consortium's Emoji Subcommittee did not make themselves available for comment before this piece went to press.)

One of the emoji's designers, Dalvi, told me that his design has been misinterpreted because of the political context into which it was released. "The timing could not have been more unfortunate," Dalvi said. "As a designer, this is the pitfall. My design can be politicized and used towards ends that I cannot control." But it's impossible to deny the Hindu-supremacist symbolism of the emoji, especially given that countless other approaches were theoretically possible. The consortium wanted to make sure the emoji was clearly a "Hindu temple," but they didn't say how to accomplish this. If Dalvi didn't intend for his emoji to be a badge of Hindu supremacism, his design choices show just how successful the RSS has been at promulgating its vision for India.

Permanent Supremacy

Use of the :hindu_temple: emoji (as its shortcode is known on Slack, GitHub, and other platforms) is still picking up, spurred on by software that suggests it whenever users type "temple," "mandir," or other words on their keyboards. The emoji has been adopted by some social media accounts dedicated to sharing temple photos and discussing temple tourism, but it has also quickly become a shorthand for expressing support for Hindu supremacy—or issuing direct threats—when included in social media usernames and messages.

The emoji has something in common with other ways in which Hindu supremacists have deployed language online to further their cause. Like the emoji, certain phrases purport to be descriptive but actually promote dangerous narratives about who is an enemy and who is a supporter of Hinduism and India. These phrases include "Love Jihad," which refers to a supposed phenomenon of Muslim men seducing Hindu girls, and "Urban Naxal," which is used to describe a certain type of English-speaking, urban-dwelling leftist thought to be sympathetic to Maoist revolutionaries operating in parts of the Indian countryside.

The central danger of the emoji is that it is as generative of people's ideas about what counts as Hindu or Indian as it is reductive of Hinduism's and India's complexities. "We have to be very careful with the process, that what we ultimately encode is going to be something that has permanence and will stand the test of time," Craig Cummings, a senior technical product manager at Amazon and the vice chair of the Unicode committee that oversees the approval of new emojis, says in *Picture Character*, a 2019 documentary about emojis. What the consortium has helped give permanence to in this case is the Rashtriya Swayamsevak Sangh's vision for Hindu supremacism in India and worldwide. It is a small but potent example of the way that Silicon Valley has chosen to make money from hate. 〰

Nayantara Ranganathan is a researcher and lawyer writing on issues of technology and society.

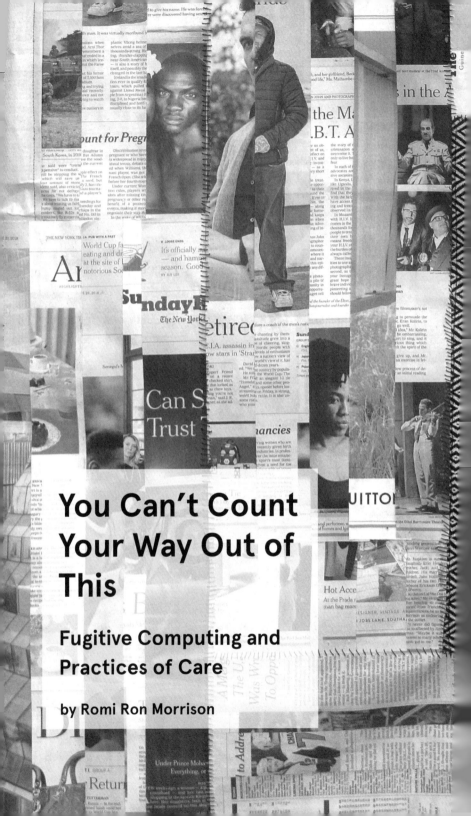

You Can't Count Your Way Out of This

Fugitive Computing and Practices of Care

by Romi Ron Morrison

This is not an act of recovery to assure you all that Black people have indeed done some shit with numbers, nor is this a fetishizing of quilts as computing objects, though they are. It is an argument for forms of computation that subtend "the beauty of black ordinary, the beauty that resides in and animates the determination to live free, the beauty that propels the experiments in living otherwise" — embedded in technologies of the living, connected to care work, to relation, to difference, and to contention.

This article is an abridged version of "Voluminous Disintegration: A Future History of Black Computational Thought," forthcoming in *Digital Humanities Quarterly*.

Excerpt from "Wayward Lives, Beautiful Experiments," Saidiya Hartman.

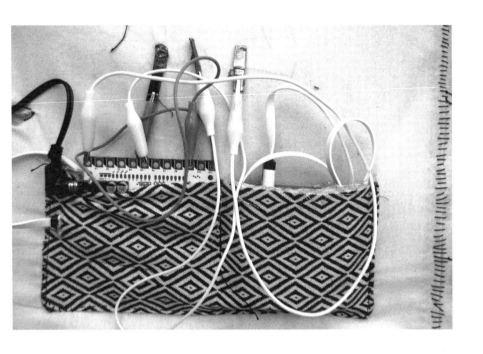

We are in a moment of radical reimagining and visceral reconnection. Caught between two worlds, we are catching glimpses of the new one in the fall of the old. Built on the longstanding rigorous work of Ruth Wilson Gilmore, Mariame Kaba, Andrea Smith, INCITE! Women of Color Against Violence, and Angela Davis, amongst many many more, known and unknown, abolition has announced itself within the popular imagination. We are seeing abolitionist organizing in the flourishing of public acts of rebellion, mutual aid projects, community accountability practices, transformative justice trainings, people gathering everyday to block evictions, calling in organizers, and hosting popular education workshops beyond the university. This organizing can also be seen in the connections made between surveillance technologies and carcerality. Increasingly, pressure is being placed on the ban of facial recognition technology as well as of big data itself.

Embedded in these calls is a politics of refusal. To refuse the grammar of the present order. To come together in all the ways that are sought to be exorcised, messy, spilling, and raucous.

Translated from Michel Foucault's term, "dispositif", in his 1977 interview "Confessions of the Flesh." The "apparatus" refers to the institutional, administrative, and physical structures through which power relations are formalized.

Based in Denise Ferreira da Silva's piece, "Toward a Black Feminist Poethic." She uses poethics to describe a speculative manner of thought to think the world differently, beyond the trappings of linear rationality that underpin Eurocentric colonization.

It is the time to ask: Why does anti-Blackness seem to perpetually overdetermine and saturate the operating system regardless of who is programming it? How do we begin to move to forms of critique and resistance that relinquish a certain focus on the apparatus and begin to disassemble the episteme, that seemingly transparent foe that perpetuates the endless production of violent techne? How might this make desirable the pursuits of situating computation elsewhere and towards a poethic of endured proximity?

Endured proximity refuses imaginations of the world in which calculable measurement is the only relationship between things. Endured proximity pushes us to meaningfully engage difference as a relationship that we are entangled within. Only by doing this will contemporary flights of liberation be imaginable.

Introduced in Foucault's text, The Order of Things, episteme comes to mean the unconscious beliefs that structure scientific knowledge in a particular time and place.

As used in Martin Heidegger's The Question Concerning Technology. Heidegger uses techne to mean a practice of revealing or bringing forth. It is connected to making not just objects but knowledge and discourse. This differs from common understandings of technology as a tool or instrument.

I look for this endured proximity through a reading of the Freedom Quilts, a clandestine system of mapping escape routes for enslaved Black people, as a vital form of computation. Quilts were constructed according to a technical protocol of sewn symbols, stitching patterns, and tied knots. This quilt code includes ten primary patterns and a number of secondary patterns. Each pattern had two meanings: signaling when to prepare to escape and giving clues to indicate safe routes of passage. After leaving the plantation, enslaved people would encounter quilts bearing single patterns left in public to air. Following these codes, they would know when to gather the tools they needed for the coming journey, the time to escape from the plantation, ways to navigate hundreds of miles to destinations in the North, how to calculate mileage between safe houses, local topography, places to find fresh clothes and shelter, and practices for recognizing other confidants.

Knowledge thrives through circulation and exchange. It is shared, encoded and stored through social relationships. Data likewise is organized to be exchanged and read by another person or machine, it also thrives through circulation and is dependent on its mode of transmission. Though data is discrete, it only abstractly represents phenomena far more continuous and entangled. It requires work to attain such distinction. Capturing, cleaning, and formatting are not natural processes but demand physical, emotional, and intellectual labor. Therefore, data is built upon embedded social relations. It is a relation between semblances of ourselves and how we engage change.

Created in quilting bees on plantations, the Freedom Quilts bring this social relationship to the forefront. Quilting bees were collective circles of primarily women and some men that sat together and constructed quilts in collaboration.

d to give his name. He was forced
er were discovered having sex

th man. It was virtually moribund b

nsions when
d. Arni Thor
remembers a
at ended in a
in which Ice-
nst the Faroe

ut his fervor
of 7,000 fans
adium.
g and trying
aid recently.
own and not
ving to watch

w outliers in

plastic Viking helme
selves amid a sea of
thousands-strong for
ing, thunder-clapping
near-South American
— is also a story of h
itself, and possibly the
changed in the last ha
    Iceland is the small
tion ever to qualify fo
team, which pulled
against Lionel Messi
ple from Argentina) in
ing, 2-0, to Nigeria her
disciplined and hard t
usually close to its far

t, and her girlfr
od life," Ms. M

N JOHN AND PH

the
.B.T

punt for Pregr

SE — GETTY IMA daughter in
ea, in 2016 But Adams
      ise the seed-
ere "treme the current
o conduct.
ping the w ople effect on
    save us The French
t of mone a seed, but
lso criticizi ly 2, has tra-

Discrimination invo
pregnant or who have
is widespread in many
sional tennis, debate o
ed when Williams, th
nant player, was not
French Open. (She wit
before her fourth-roun
    Under current Wom

e us ob-   the
ht of us,   cri
eflect on   ove
I.V. and   onl
: to con-   fea
— as a   l
ery short   ad
             siv
le treat-   l
le oppor-   lik

Because these quilts were often sourced from various spare pieces of fabric, they were piecemealed and required numerous people to collect and plan each quilt. Quilting bees were social spaces and sites of convergence, as they were vital stops along the plantation grapevine.

Quilting bees became important gatherings where enslaved people "were able to compile facts regarding geography, landmarks, places to avoid, obscure trails, mileage, and the locations of safe places where food and rest were waiting... Many escaping slaves knew where to go and how to get there. Former runaways shared their own tactics and routes of escape. Most early escape attempts were individual efforts by slaves, not part of any organized cooperative ventures headed by Northern abolitionists."

J.Tobin, R. Dobard, *Hidden in Plain View: A Secret Story of Quilts and the Underground Railroad*. (2000). First Anchor Books, New York, NY, p 74.

As Tobin and Dobard document, sites such as quilting bees became gathering spaces through which communication between free Black people in the North, white abolitionists, and the enslaved took place across numerous plantations and regions. Black peoples before capture, from various geographic, historical, and cultural sites, were not fixed in their ultimate difference upon which the only mediation could be found through measurement. Instead, enslaved Africans held onto their cultural memories and combined them with others stolen from their lands to create new creolized, semiotic systems. This meant working within the entanglements of difference to communicate with each other, circulate knowledge, and to build wholly different cultural systems within the diaspora. This is how the Freedom Quilts were made, as a vital form of Black computational practice that forces us to rethink what computing can be, when freed from its dependence on colonial pursuits of managing bodies, spaces, and resources.

When we carve out time and space to pay homage to an aspect of life, we are engaging in a ritual. *Rituals of Black Fugitivity: Protection* is a mixed media sound installation that includes a large interactive quilt made of repurposed newspaper, fabric, and conductive thread. The quilt is stitched from numerous rectangular strips of overlapping and fragmented newsprint in the symbol of the Log Cabin. Within the fugitive practices of the Freedom Quilts, the Log Cabin was a code used to signify a space of sanctuary, or rest, within the violent landscape of Plantation Capitalism. *Rituals* stages a scene of stolen time by which Black people carve (im)possible moments of freedom through ritual practice. This scene falls out of linear progressive time and compresses space-time, folding the past into the future, where ritual meets technology, and thread meets circuit. At the center of the quilt are three folded paper fans triggered by touch that play a series of sound compositions detailing my own rituals for daily protection.

By using the term "Plantation Capitalism," I am referencing Clyde Woods' work in *Development Arrested: The Blues and Plantation Power in the Mississippi Delta*. Here, Woods argues that the Atlantic Slave Trade is a pillar for the origins of capitalism due to the totalizing control over land, labor, people, finance, and information flow that the plantation gave rise to.

Sound works as an invitation, and voice finds dialogue within the interface of the quilt. By listening to my rituals for protection it asks you to consider your own. This installation is a database that requires us to step into it, it is a space, a carved respite. It is not important if our experiences are commensurable or equivalent, they need not be measured, the interoperability of the exchange is partial; what is important is that we are engaged in a relation. The datum is not separate from my body, or the accrued practices of imagination and rebellion that my ancestors continue to carry. This relation is one that we contend with through maintaining an endured proximity outside of measure. This means that as we measure and collect data about the external world, we cannot continue to think of ourselves as transparent or inconsequential in our relationship to what is being indexed. Instead we must contend with our relationship to the index, to see ourselves in the Black spaces of its stolen time, and in relation to fugitive knowledge. It suspends the question of technology from being answered, and leaves open room for ritual techne to wander.

The question of technology: technology is suspended, or left open as a question without a definitive answer.

Here the digital is returned to the digit, to the hand, to haptics, and texture, and textiles, to the process of making through material and proximal relations to each other, to making through endured engagements with difference. My engagements with fugitive computation are indebted to understanding this natural disorder, this turn to ensuring a future through fugitive acts taken now. To move beyond the paralysis of precarity, preparing a new world in the shadows of the old.

Building Better Technology Together

Audrey Eschright on Open Source and Organizing in Tech

Audrey Eschright has been starting and stewarding open source projects, conferences, and communities in Portland for over a decade. Open Source Bridge, which she and other volunteers ran for ten years from 2008, was "intended as a call to action to become better citizens, by sharing our knowledge with each other." (Last year, she wrote that if she were doing it all over, she would ditch "citizens" in favor of something more expansive.) In 2015, she founded a "feminist hacker zine" called *The Recompiler*. These and Eschright's other projects all point to a version of open source that's scrappy, place-based, accessible, and motivated by something like a sense of civic responsibility. If you still believed in such a thing in 2020, you might call it small "d" democratic.

There are other versions of open source: technically impressive but interpersonally abusive, or the commercially beloved version that extols the virtues of making uncompensated technical contributions to codebases owned by corporations. Eschright has been taking the best parts of them and leaving the rest. And like any good feminist, she has been engaging with questions of labor. Eschright has found that her experience building open source project infrastructure has translated easily to building solidarity infrastructure with other tech workers.

We rescheduled our first conversation due to disruption caused by the record-breakingly destructive wildfire season that blanketed the Pacific Northwest in hazardous smoke for three weeks in September of 2020. When we finally caught up to talk about what open source can and should be, both of our cities had been dubbed anarchist jurisdictions by Trump's Department of Justice.

————

How did you first get involved in open source software, and what drew you to it?

Before there was open source, there was free software. I generally had this late-1990s, early-2000s understanding that, within the technology world, there was free software, which was often literally free, and then there was Microsoft stuff that cost a lot of money. I grew up poor and it was financially very difficult getting through college, so just in terms of access, I thought, "This Linux thing that I put together from various parts? This is good." I went to the University of Washington in Seattle. Microsoft land. My experiences were shaped by both my economic status and the ways I saw Microsoft impact the university. It all funneled me to the free software, and, later, the open source movement.

I got into the community side of open source through volunteering at Free Geek for a couple of years, which is a computer recycling facility and educational institution. They have a great founding story about starting the organization after seeing a computer monitor that someone had thrown into the Willamette River.

Also, they give you a computer that runs Linux if you volunteer there. I graduated from college during a recession, so it took me a long time to find work and my computer died in the middle of that process. I figured I'd put in some volunteer time and get a computer. That's what brought me into this local community of user groups, where people were teaching each other about programming. Most people would call these "meetups" now, but the user groups hosted at Free Geek tended to focus on a particular technology, like Perl or PostgreSQL. That community supported me getting into this career. So, it made sense for me to find ways to support other people too.

At some point, you left Seattle and went down to Portland. You got involved in the Portland open source community in a big way, planning multiple conferences over the course of a decade. What drew you to that work?

I just kept asking questions. The question that brought me to Free Geek was: How do I get a computer? The question that brought me to user groups was: How do I learn how to write software for it? And then: How do we cross-pollinate the different user groups? The answer was to host meetups and organize conferences. How do we keep track of all the events we're hosting and resources we're sharing? In 2008, I ended up building Calagator, which is an open source calendar app that's also kind of a wiki. I'm still the maintainer.

It's always been like that: We need this thing. How do we get it? How do we make it? Where does that take us next?

For the past decade, it's always led you back to open source. I'm curious what your thoughts are on the state of open source today and the phases you've seen it go through.

Open source is not one big thing; it is a giant umbrella for many different activities and goals. For a long time, Linus Torvalds said that the goal of the Linux project was technical excellence. The technical excellence perspective has meant that open source is the base layer of most technologies that we work with. Linux is in everything. Developer tools, web servers, our programming languages are all open source or built on open source components. Very little of our technical stack is proprietary. That's part of what drew corporations to open source. That's also what drew me to open source: it provides low-cost solutions for people.

> *"I do not care about code quality or how fast the process is, or whether it scales. Where we need to innovate today is around community safety."*

What has become *more* complicated over time is the ideological piece. I've written about reframing open source, so that it values the people who make it more than anything else. The lens we don't apply nearly enough is: How do we build technology in a way that is beneficial to people? How do we make software for what we actually need? Software that supports us, teaches us, builds communities, and solves real problems. We need to talk about this from a community perspective, from a human perspective.

Technical excellence cannot be the primary goal. I do not care about code quality, or how fast the process is, or whether it scales. Where we need to innovate today is around community safety. Many of the technical challenges that we work on as developers have had solutions for decades. I think that we can take what has worked for us from that version of open source and throw the rest of it back out there.

My answer would have been much tidier a year-and-a-half ago, but my perspective has shifted from thinking about the open source community as the center to thinking about where open source sits within the rest of the world. It doesn't give me as easy of an answer.

Mind the Gap

You founded *The Recompiler* in 2015 to feature writing on technical topics like DNS and floating point numbers. The website describes the magazine as a feminist hacker zine, whose tagline is "Technology for the world we want to live in." How does *The Recompiler* and the larger Recompiler Media project fit into that?

I have to be honest here and say that the question *The Recompiler* was answering for me was not: How do we build better technology together? Although that is a great, true mission statement. The question for me was really: How do I continue to work in technology—after a certain point in my career, knowing what doesn't work, knowing that this industry continues to marginalize people and exacerbate power differentials?

Is there a specific experience you were thinking of? It sounds like a shitty workplace.

It was, but it was also burnout. It was the mid-career developer experience of someone who's not a white man. So many conversations I've had are about that experience, and they're almost always framed as: Do we stay or do we leave? And I thought, well, what else, what's another option? That's really how *The Recompiler* came to be. Prominent writers and speakers in the industry kept effectively saying that there were no women in technology. But I knew women in technology, I knew people of color in technology, I knew people from lots of marginalized groups in technology. So the Recompiler projects were about creating a platform to reflect that we already have technical experts representing a lot of different kinds of people.

Across the Recompiler Media projects—the newsletter, the podcast, the magazine, the blog—you cover a wide variety of subjects: technical deep dives, police brutality in Portland, how surveillance technologies work at the code level. How do you see those all fitting under the same umbrella?

It goes back to asking questions and following them, whether it's operating systems or legacy software or what the police are doing in Portland. I always wanted the deep dives to be what distinguished us. Really good, easy-to-understand explanations of technical topics that everybody bullshits about.

You're interested in demystifying.

Yeah. We need to understand how things work for ourselves. And, if you're asking questions about how technology impacts people's lives, policing and surveillance keep coming up. Palantir, facial recognition, who had access to what databases after the 2016 election—there was no way to avoid those topics. So, that's what brought the Portland protest blog into *The Recompiler*'s wheelhouse.

Can you talk about the organizational structures of *The Recompiler* and how you find contributors and collaborators?

I usually give my title as either publisher or editor-in-chief, depending on who I'm talking to. Basically, I pay the contributors. The only person who does not generally get paid is me. I would love to be doing this full-time, but this is why I've come back to having a day job. I did do *The Recompiler* full-time for a couple of years, but funding-wise it's better for me to have a software engineer job.

That is also part of *Logic*'s funding model.

It helps. In terms of finding contributors, we've done an open call for every issue. I encourage people to put stuff in and people come to me asking if their idea would be a good fit. That's how we ended up with the *Responsible Communication Style Guide*. My friend Thursday Bram had always wanted to put out a guide like that. I thought it was a good idea and that people would use it, so we did it.

> "*What do we need that nobody is doing? I'm always thinking, what if I grab another person and we make something to fill the gap?*"

With the newsletter, the person who had done it originally wanted to step away because talking so much about all the ways that technology harms people got to be really draining. I started asking around and somebody introduced me to Margaret Killjoy, who does the newsletter now. She makes most of the decisions on that, and she's brought a more leftist, anarchist focus, which I think is very appropriate for the moment.

Since first reading Margaret in the newsletter, we've been listening to her anarchist prepper podcast in our house. Did you

ever think that *The Recompiler* would be one hop away from an anarchist prepper podcast?

One of the skills that I bring is that I'm always looking for the gap—what do we need that nobody is doing? And I'm always thinking, "What if I grab another person and we make something to fill the gap?" That's how all these projects start, in this fractal way, where we're going along and then I see an opportunity halfway through and branch off. What I hope is that we end up with a whole organism of projects that are supporting the different interlinked needs we have.

> *"We became educated about a lot of things, and then we told other people, and then they still didn't do what we needed them to do."*

No Bosses, No Knife Missiles

I want to switch gears and ask you about labor organizing and union busting at NPM. I found a GitHub gist that you published in 2015, titled "A Feminist Hacker's Guide to Labor Organizing," a year or so before Trump's election sparked the current wave of white-collar tech worker organizing. In 2015, it was a national press story that Googlers were anonymously sharing their salary data in a spreadsheet, and #TechWontBuildIt was still years away. Do you remember making that gist?

Wow, yeah, I do remember that. That was after a conversation I was having with someone in a user group. I wrote it down, sent it to the list for the group, and haven't looked at it in years. I wonder if there's anything I would disagree with now.

It's pretty solid! Can you talk about why you were thinking about tech worker organizing in 2015?

Before people were talking about Trump and how technology would be used under his administration, I was concerned about power differentials within tech: the ways that tech companies drive developers to burn out; the fact that Silicon Valley companies saw us in Portland as a source of cheaper labor; the fact that you can't make your boss stop being racist, but you can create consequences for your boss being racist.

I wrote that gist because we were talking in the user group about all the things that hadn't helped us. The employee resource groups that companies created to try to make us feel heard, HR, all these trainings that were like, "Here's your bias," and then everyone says, "Yup," because bias does not go away when you learn about the existence of the bias.

There was this real frustration about how things weren't changing. We had been in the industry long enough to have seen some efforts that went nowhere, and not because we didn't try very hard. We became educated about a lot of things, and then we told other people, and then they still didn't do what we needed them to do. I just got to this place where I didn't want to keep nicely asking: "Please stop being racist, please stop being sexist." I wanted to do something that we hadn't tried yet.

Labor organizing is a solution. It helps us leverage what we have, which is people—multiples of people—against these institutions. But it requires that workers are informed and talking to each other. In many ways, it's like organizing an open source project: we have people, we have a need, how do we share what we know with each other, how do we build something together?

How did that start at NPM?

My organizing at NPM happened by accident. Right before I joined, the company brought in this CEO who was the stereotypical guy that you hire so that the company can finally make money for the investors. But the company had attracted people whose ideals were at odds with just cashing out at any cost. There had also been unaddressed burnout issues before the CEO joined, and he made them worse because he was so terrible to the people who had put in so much work to get NPM to that point. Yet we were encouraged to talk about it all: about burnout and retention and the company's new focus.

I was in a team meeting at the company all-hands, which was an offsite, but I was one of two people who was there over video. People were saying things like, "Well, you know, if we had HR, then we'd have somebody to help with some of these management issues. And if we could get them to do this, then we'd have that." And I was like, "That's cool, but we could also unionize."

I was having this really awkward experience over video, I was cranky, and, as I mentioned, none of my previous experiences led me to believe that we were going to be able to make incremental institutional change in the way that some people wanted. so I didn't think that it was a particularly outrageous thing to say. I thought some people would agree with me, although I did have a moment of doubt afterwards when it occurred to me that maybe everyone thought that was a terrible idea and there was nothing I could do about it because they were in a room together and I was just there on a screen. But, in fact, there were people who were thinking, "That's what I wanted to say. She just said it."

Were you friends outside of work already? Because I'm guessing you weren't doing this in the work Slack.

After that conversation, two of my coworkers were trading phone numbers and talking about meeting up later. So over the Zoom call, while the bosses were out of the room, we all exchanged numbers. Our initial organizing group came out of that.

Now that everyone is a remote worker, no one has those in-person opportunities. But what you can do is say, "Hey, I'd like to talk to you about how things are going." Companies all have their own ways for workers to connect with each other. And then over Zoom you say, "A few of us are talking about our concerns in this other place—would you like to join us? How are you feeling about what the CEO said or that policy change or whatever?" Just ask. And then you get off the Slack. No Slack DMs about how you're going to unionize!

Once we had our organizing group at NPM, people started doing something that's really common in these groups. They'd say, "This is really bothering me." And then other people chime in and say, "Yeah, that really bothers me too." Or, "I think I'm getting underpaid. What are you getting paid?" Or one person voices concerns about the lack of trans healthcare, and then someone else would offer to ask about it for them. That kind of thing often happens automatically. I'm not just talking about NPM here. But people start having these conversations about what is actually happening and agreeing with each other. Even if it's just a solidarity forum, that's valuable.

What really grinds us down in tech is feeling like we're not heard or seen. We're told that everything is fine, that everyone loves it here! The solidarity groups can be validating. And people are motivated and impacted by the kinds of topics that come up there, so they want to teach each other—about forms of marginalization or using inclusive language or organizing in general.

Something that's been interesting is that when I asked people what brought them to the organizing group, they often responded that they had some experience of organizing from another area of their lives: they grew up in a union household, they organized in college, they had a family member who talked about organizing—more tech workers than you would think. And that's a stepping stone for them.

At some point, you were illegally fired from NPM.

Rather quickly, yes. Because I made that comment in front of my whole department and it didn't take long to make its way back to the CEO. They didn't know who all was involved, but they got rid of who they thought the troublemakers were. That ended up being a sizable part of their open source expertise.

One of the other people who was fired said that they were going to file a National Labor Relations Board complaint. I'd never even thought about it. I just had to find another job immediately because I was coming out of doing *The Recompiler* full-time, so I didn't have any buffer left. But this person wanted to try so I said, "Okay, let's try."

The combination of the firing and the NLRB settlement negotiation was honestly one of the most stressful things I've ever gone through. I was job searching and then starting a new job and worrying that I was going to spend my fortieth birthday testifying. There were so many calls about what we were going to ask for and what we'd accept and what was happening with the negotiations. I'd be on the phone wandering through Muji, totally overwhelmed, staring idly at a houseplant. And NPM's CEO was a real dick to us. When we thought we had agreed to a settlement to the NLRB complaint, he reneged on the agreement.

I learned a lot, like that I don't ever want to go through that again without a union filing the complaint for me, which is how it usually happens. But the Oakland NLRB office was great.

I read in some of the reporting that NPM management eventually settled with three of the workers who were fired and who weren't managers and that, as part of the settlement, the NLRB made the company remind all the workers that they are allowed to organize.

Yup. Reminding people of their rights turns out to be one of the stock things the NLRB does in these kinds of settlements. We were really into it.

We also received back pay for the period between our firing and the settlement, and we received a few weeks of additional pay that was similar to the severance agreements we turned down in order to be able to talk about what happened.

> "*They grew up in a union household, they organized in college, they had a family member who talked about organizing—more tech workers than you would think.*"

As we wrap up, can you talk about where you're at politically now and how you see that as interwoven or not with your professional work as an open source developer?

We are living through an extreme moment in history. The things that felt relevant even a couple of years ago don't have the same relevance or urgency now. I was focused last year on different kinds of ethics clauses in open source licenses. I'm not doing that right now. I've spent my summer watching people be beaten by the police. I've heard politicians say again and again that tear gas is bad, but then there is still tear gas. I'm one of many people who have been radicalized by this summer in Portland—by what the police and the mayor have done. I was already a lefty, but it's refocused things for me.

So I'm still thinking about labor and open source and access in tech, but there are some life or death issues in front of us that software is not going to solve. The technology industry has landed at this point where it is not separable from prisons, policing, and surveillance. If companies are there to make the profits, do the launch, and get the return on investment, they are going to be doing this work. Look at policing as a budget line item in our cities, even small cities. Look at the military contracts, the intelligence contracts. That is where the money is. Companies say, "Oh, well, we don't work with ICE; we only work with DHS!" Or "We work with Raytheon, but not on the knife missiles!"

We have to get rid of ICE. If we don't want Palantir helping DHS identify immigrants to put in cages, we have to get rid of Palantir *and* DHS. We have to abolish these systems and, when we do, the tech industry will have to find a way to make money some other way. ⁓

Specter in the Machine

by Evan Malmgren

In the 1990s, as the internet was being turned into a shopping mall, a group of radicals built a digital commune.

————

Pit Schultz was sitting in a Kreuzberg art gallery, less than a mile from the ruins of the Berlin Wall, when he sent one of the first emails to a mailing list that he had just helped launch. Schultz laid out a vision for what it might become:

> *It should be a temporary experiment to continue the process of a collective construction of a sound and rhythm — the songlines — of something we are hardly working on, to inform each other about ongoing or future events, local activities, certain commentaries, distributing and filtering textes, manifestos, hotlists, bits and blitzmails related to cultural politics on the net. It's also an experiment in collaborative writing and developing strategies of group work... The list is not moderated. Take care.*

It was June 1995, and the internet was changing in fundamental ways. The US government–funded National Science

Foundation Network (NSFNET), once the backbone of the internet, had been decommissioned a few months earlier. New companies like Amazon, Yahoo!, and Netscape were racing to cash in on an early wave of commercialization, and a rising priesthood of techno-utopians gathered around *Wired* magazine—launched in San Francisco in 1993—to herald the coming digital economy as the harbinger of a more unified, democratic, and horizontal world.

The 1990s were a heady time, but not everyone was convinced. If *Wired* was the pulpit for a new gospel of venture-funded tech, Schultz's mailing list, called Nettime, was an effort to build a home for the early commercial internet's discontents. Drawing on various contemporary anticapitalist currents, from the anti-globalization movement to Italian autonomism to Berlin's lively squatter movement, Nettime aimed to synthesize an alternative to the techno-determinist optimism oozing out of Silicon Valley: the worldview that media theorist and Nettime regular Richard Barbrook named the "Californian Ideology" in a 1995 essay co-written with Andy Cameron.

The Californian Ideology, through its "bizarre mish-mash of hippie anarchism and economic liberalism," celebrated the rampant commodification of digital networks as a force for personal liberation. The 1990s are often remembered as a time in which this vision of the internet went unchallenged, but the Nettime crowd wanted to chart a different path forward. They developed theories of digital culture, pioneered tactics for new media activism, and wrote ground-zero critiques of the commercial internet as it took shape around them. The mailing list itself was also a platform for experimental forms of collaborative writing that tried to embody a different *experience* of being online. At a moment in history when profiteers and privatizers were terra-forming the internet into the market-saturated system we know today, the Nettime circle gestured toward a more collective, less commodified alternative—but only vaguely.

A Collective Undertaking to Deconstruct the Utopian *Wired* Agenda

Nettime lived on the internet, but it came from Europe. The idea for the mailing list was first proposed by Schultz and his friend Geert Lovink at a meeting of the *Medien Zentral Kommittee*—German for "Central Media Committee"—at the Venice Biennale. The Kommittee was a loose collective of academics, net artists, and new media activists who believed in the necessity of creating alternatives to both the Silicon Valley worldview and the commercial internet it had inspired. In doing so, they drew on materials closer to home, building an umbrella for an anarchic strain of critical net culture that flourished across Europe.

> "*The 1990s are often remembered as a time in which this vision of the internet went unchallenged, but the Nettime crowd wanted to chart a different path forward.*"

Whereas *Wired* drew from the ranks of a thriving private tech sector, Nettime's milieu came from a relatively communal and localized hacker culture. This took shape, for example, around spaces like c-base, the member-funded nucleus of Germany's rapidly expanding hackerspace scene, which went on to promote free public internet access via wireless networks. Another spiritual progenitor was Austria's Public Netbase, a nonprofit internet service provider and new media initiative that promoted "network democracy from below," openly clashing with the country's right-wing government.

Europe had also been home to a number of ambitious publicly funded efforts to extend social and civic life into the digital realm. Perhaps the most famous was France's publicly owned Minitel network. Rolled out in the 1980s, Minitel was the most successful online service prior to the modern internet. It ran across nationalized phone lines, distributed free terminals, and boasted a peak of twenty-five million users out of a total population of sixty million.

By the 1990s, similar initiatives had cropped up around Europe at the municipal level, like Amsterdam's Digital City (DDS). A citywide free-net, DDS grew out of the pirate radio scene and aimed to create a universally accessible network that could guarantee certain basic rights online. There were public terminals, and anyone could sign up for a free account with email, internet access, and space for a homepage. Instead of being organized around the free market, the network's architecture was designed with the city metaphor in mind: you received mail at the "post office," links were accessed through a "station," "public squares" hosted government services, organizations and companies could rent "shops," and the entire system was navigable through a graphical city interface.

Public networks like Minitel and DDS can help us understand why European hacker culture diverged so sharply from the cult of Silicon Valley. While the US obscured the internet's publicly funded origins behind a veil of bootstraps entrepreneurialism, many Europeans first encountered mass computer networks as explicitly *public* entities.

Nettime took inspiration from these projects, and attracted many of their architects, but the list itself represented a different kind of intervention. If projects like c-base, Public Netbase, and DDS explored new ways of creating and organizing networks, Nettime was also an attempt to theorize and embody a new way of *experiencing* them. In the words of Geert Lovink—a cofounder of both Nettime and Amsterdam's DDS—the list was "a collective undertaking to deconstruct the utopian *Wired* agenda. Not directly, in word or academic texts, but by doing."

> *"While the US obscured the internet's publicly funded origins behind a veil of bootstraps entrepreneurialism, many Europeans first encountered mass computer networks as explicitly* **public** *entities."*

In the early to mid-1990s, the internet could be a disorienting place; a seemingly endless labyrinth of what *Baffler* contributor Kate Wagner described as "haphazardly designed, amateur-generated sites." But it also often felt more grounded and neighborly than today's internet. Most online communities were small and focused enough for participants to develop personal familiarities, shared norms, and something like a localized collective consciousness—"netiquette," as it was called. In a time before platform capitalists had carved the net into siloed empires of attention-time, it was still possible for anti-capitalists of the sort that gathered around Nettime to see the internet as a catalyst to dissolve commercialization and competitive individualism—but by 1995, things were starting to change. Developments like online shopping, powerful search engines, and interactive advertisements were beginning to rend the federated collectivism of the early internet into a world of quantitative efficiency and algorithmically mediated "users." A window of possibility seemed to be closing. Nettime was determined to keep it open.

Living in Social Time

Nettime's name was chosen as an alternative to "cyberspace," the dominant metaphor for understanding the internet in the '90s. "Cyberspace" renders the internet in spatial terms, and evokes images of highways, libraries, webs, clouds, and shopping malls. All of these tend to naturalize concepts like scarcity and enclosure, which in turn lend themselves to the possibilities of exclusive ownership, exploitation, debt, or rent.

> *"The original vision was to initiate something like a perpetual conversation, without editors, boards, gatekeepers, or centralized moderation."*

By contrast, "nettime" renders the internet temporally. Whereas the concept of a spatial network frames humans as occupants of a fixed virtual world—one that could be chopped up into shopping malls—"nettime" suggests that their mutual engagement fundamentally constitutes the network itself—that there is no network without the nodes it connects. Rather than passively "going" online and browsing shelves, we *actively produce* the network together, in real time, through our collective participation. "The time of nettime is a social time," wrote Pit Schultz in the introduction to an October 1996 Nettime publication. "Time on the net consists of different speeds, computers, humans, software, and bandwidth, the only way to see a continuity of time on the net is to see it as an asynchronous network of synchronized time zones."

In trying to embody this "social time," Nettime pioneered a practice of "collaborative text filtering," a continual, self-organizing process whereby texts were submitted to the broader group, replied to, expanded upon, and ultimately flowed into a collective train of thought. The original vision was to initiate something like a perpetual conversation, without editors, boards, gatekeepers, or centralized moderation; the list's "filter" was the collective interests and capacities of its self-selecting membership. In the introduction to *ReadMe!*, a collectively edited book of Nettime essays published by Autonomedia in 1999, the list is described as "always different from what it was a moment ago; it's always discovering something new about itself. As such, it is a working implementation of what subjectivity might become in an online environment."

In practical terms, this meant that Nettime served numerous functions: it was a tactical bulletin for Europe's anarchic hacker community; an open source prepublication platform for academics; and a forum to discuss current events, announce events, post manifestos, and theorize the commercial internet as it came into focus. Prominent works to come out of Nettime were occasionally circulated at conferences, republished in magazines, or rounded up into physical publications—but online, the emphasis was always on a perpetual process of becoming. Whereas early online messaging programs like AIM (AOL Instant Messenger) and early virtual communities like The WELL (Whole Earth 'Lectronic Link) operated on the concept of exchanging content between discrete, static users, hierarchically sorting information, and publishing text in a self-contained, finalized state, an unmoderated mailing list leaves participants with no common structure or interface but one another.

Collaborative text filtering was meant to recover, in a many-to-many communication system, the experience of temporally interconnected thought that is foundational to real-time conversation. Nettime members developed a collective style of writing that embraced its own incompleteness, questioning the notion that any text can or should be understood as finished, or that any thought is worth isolating from its discursive context. This practice generated prescient missives and essays on applying copyleft practices to non-software intellectual property, the reality that the internet was already doing more to restructure labor rather than expand leisure, the creeping power of data collection and surveillance, and the creation of a new class of "information proletariat." Above all, it emphasized the network's participants as its creators, rather than mere residents.

Haunted Variables

Nettime fought for alternatives to the privatized net, pushed back against the Californian Ideology, and experimented with a collective experience of being online. But the list had considerably less success in articulating a comprehensive alternative to the commercial internet at scale, or bringing its experience to the broader public. The mailing list grew to thousands of subscribers, but its vision for an alternative to the commercial internet remained dormant, and its members ultimately failed to alter the course of the internet's commodification.

This failure was partially rooted in a vagueness at the heart of Nettime's challenge to the *Wired* line. While mailing list participants could momentarily *enact* a more collective experience through collaborative text filtering, the lack of organization also ensured that no consensus around a comprehensive alternative would emerge. Nettime participants tended to share certain anti-capitalist principles—a preference for publicly owned internet infrastructure, the decommodification of intellectual property, and the abolition of various digital hierarchies—but the particulars were subject to endless debate. Real-world implementation was thus limited to aesthetic interventions—culture jamming, *détournement*, and critical net art—and local, uncoordinated, long since defunct projects. Media theorist McKenzie Wark later observed that Nettime was united by "a negative consensus around the need for a countervailing theory." That negative consensus never evolved into a positive one, because a single countervailing theory never emerged.

Attempts to build such a unified theory tended to fall flat, as when Richard Barbrook sent an essay called "Cyber-Communism" to the list in 1999. "A spectre is haunting the Net," it began, "the spectre of communism." Barbrook's central contention was that everyday internet users were already on a path to transcending the profit-driven logics of the privatized net, thanks to the popularity of gift economies around phenomena like user-generated content, open source software, and peer-to-peer networking. In other words, he argued that the experiments in collaborative text filtering that Nettime was undertaking to cultivate a more collective experience of the internet were already being superseded by mainstream initiatives with far greater reach.

Twenty years later, a commercial internet dominated by a handful of platform oligarchs makes Barbrook's optimism easy to dismiss. But it didn't take the power of hindsight to notice that his "wait and see" determinism was functionally indistinguishable from the techno-utopian *Wired* set. A satirical Nettime response to Barbrook's essay captured the point:

> *Subject: THE GIF ECONOMY: How Several Layers of Lossy Images Are Synthesized into a Moving Image that Will Animate the Masses and Inspire Them to Do What They're Doing Anyway, Namely, Clicking Their Way to Liberation; Or, How I Learned To Stop Worrying and Love the Californian Ideology*
>
> *<-----! DECLARE ALL VARIABLES !----->*
>
> *HAUNTED_VARIABLE = (EUROPE | WORLD | NET)*
>
> *<-----! END DECLARE ALL VARIABLES !----->*
>
> *Just imagine ... a specter is haunting HAUNTED_ VARIABLE for the second time. Luckily, if it's a tragedy the first time, the second time around it's only a farce—so fear not: no more messy dogmatic truths, state bureaucracies, apparatuses of oppression, proxy wars, or national collapses. CyberCommunism[TM] GUARANTEES you won't have to change a single setting, preference, or property in order to build a communist society! You can contribute DIRECTLY to the construction of a workers' paradise on a GLOBAL scale from the comfort and privacy of your own home or office WHENEVER you feel like it—just by surfing the Internet!*

Largely derided on Nettime, "Cyber-Communism" nonetheless displayed a foundational shortcoming of critical '90s net culture: an aversion to the firm commitments and big-picture thinking that would have been essential for mounting a serious challenge to the internet's rapid privatization. One of Nettime's signature critiques of the *Wired* line was that the techno-utopian gospel's apparent optimism about the networked future disguised a fundamental pessimism about the role that humanity might play in it. Humanity was not the subject of this future; technology, mediated by the profit motive, was. Utopia was coming, whether we liked it or not.

Yet Nettime's participants often reiterated this same idea in a slightly different register. In the end, Barbrook shared the techno-utopian faith that, through the commercial internet, humanity was automatically generating a better future, albeit a communist rather than a capitalist one. Lacking a coherent political program, such immanentist arguments amounted to little more than a vague hope that things would work out in the end or, at best, a collection of abstract demands: "Deprivatize corporate content, liberate the virtual enclosures, and storm the content castles!"

Echoes of Electric Agora

By 2020, a handful of Nettime's once-utopian ideas have become ubiquitous facts of the commercial web—but only in forms that have been disfigured to the point of unrecognizability. Collaborative text filtering, for example, has become a core fea-ture of how modern social media platforms create and manage feeds. Rather than employing formal gatekeepers or editorial staff, websites like Facebook, Twitter, and Google rely on collec-tive behavior and decentralized editorial discretion to produce evolving, personalized hierarchies. But most of the agency has been transferred from conscious human actors to algorithms, with networks operating beyond the scale of comprehension.

Meanwhile, Nettime was forced to abandon its anarchic commitment to total decentralization around the turn of the millennium, and began tolerating light moderation once the list got big enough to encounter a few deliberate disruptors—as well as a general tendency for men, in particular, to use it "to compare the length of their bookshelves," as communications professor Matthew Fuller put it in 1998. Given that decentralized collaborative text filtering proved unmanageable at scale, and given the extent to which corporate social media platforms have adapted the same techniques to alienate "users" from their own experiences of the internet, it may be best to understand Nettime's experiments in collective subjectivity as windows into a time when the internet's corporate trajectory seemed less inevitable, rather than practical steps toward building a decommodified internet from within. Like the broader anti-globalization movement that influenced it, the Nettime circle was ultimately naïve about how easily the system would reabsorb aesthetic transgressions and hyperlocalized struggles.

Nettime has receded into obscurity since its halcyon years of 1995–2001, but the list is still running and open to new members after twenty-five years. Some of Nettime's early disciples have even risen to positions of real influence, buoyed by the post-2016 techlash and the resurgent popularity of democratic socialism. Richard Barbrook, for example, coordinated Jeremy Corbyn's 2016 "Digital Democracy Manifesto," which proposed significant state support for platform cooperatives, open source software, and public broadband expansion—structural solutions that harken back to the pre-commercial European internet. As Nettime's critiques bubble back to the surface amidst a renewed climate of tech skepticism, the list's archive provides lessons, warnings, and a usable intellectual history for today's tech-skeptical left.

The Nettime circle may have failed to stall the rapid privatization of the internet or dismantle the Californian Ideology in the '90s, but today, these forces face far more mainstream scrutiny. As an ascendant techlash resurfaces many of the European hacker culture's early critiques of the private internet, and looks for tools to build an alternative, the list's archive reveals a forgotten forerunner that kept a torch burning in dark times. ∿∿

Evan Malmgren is a research associate at Type Media Center who has written about power and communications technology for *The Baffler*, *Dissent*, *The Nation*, and others.

Decelerate Now

by Gavin Mueller

A potent strain of Luddism runs through two centuries of workers' movements. It's time to reclaim it.

———————

Adapted from *Breaking Things at Work: The Luddites Are Right About Why You Hate Your Job* by Gavin Mueller. Copyright © 2021 by Verso Books.

The original Luddites—a movement of early nineteenth-century English weavers, who infamously smashed the new machines that transformed a skilled and well remunerated livelihood into low-grade piecework performed by children—did not oppose technology in its entirety. Indeed, as skilled craftspeople, they were adept users of it. Rather, they fought against what they referred to as "Machinery hurtful to Commonality," which sought to break up the autonomy and social power that under-pinned entire vibrant communities, so that a new class of factory owners might benefit.

With every gig mill and stocking frame wrecked in the night, they identified not only their enemies, but their allies, forging new practices of solidarity. By targeting technology, they politicized it, revealing new inventions as what Karl Marx would later describe as capital's "weapons against working class revolt." And in this revelation, another: an alternative vision of how work and technology might be organized, according to what the Marxist craftsman William Morris later referred to as "worthy work," which "carries with it the hope of pleasure in rest, the hope of the pleasure in our using what it makes, and the hope of pleasure in our daily creative skill."

Many subsequent workers' movements have had a Luddish bent: they understood new machines as weapons wielded against them in their struggles for a better life, and treated them as such. But intellectuals on both sides of the class struggle have often characterized the Luddish perspective as shortsightedness, or downright irrationality. In spite of their political commitments to the working class, Marxist theoreticians have often seen the capitalist development of technology as a means for creating both abundance and leisure, which will be realized once the masses finally take the reins of government and industry.

In order to create a successful radical politics, however, Marxists must become Luddites. That is, the radical Left can and should put forth a decelerationist politics: a politics of slowing down change, undermining technological "progress," and limiting capital's rapacity, while developing organization and cultivating militancy. Letting Walmart or Amazon swallow the globe not only entrenches exploitative models of production and distribution; it channels resources to reactionary billionaires, who use their wealth to further undermine the relative position of workers by funding conservative causes like tax cuts, school privatization, and opposition to gay marriage. Letting technology take its course will lead not to egalitarian outcomes, but authoritarian ones, as the ultra-wealthy expend their resources on shielding themselves from any accountability to the rest of us: post-apocalyptic bunkers, militarized yachts, private islands, and even escapes to outer space.

> *"They understood new machines as weapons wielded against them in their struggles for a better life, and treated them as such."*

Decelerationist politics is not the same as the "slow lifestyle" politics popular among segments of the better-off. The argument for deceleration is not based on satisfying nature, human or otherwise, but in recognizing the challenges facing strategies for working class organization. The constant churn of recomposition and reorganization, which media scholar Nick Dyer-Witheford calls "the digital vortex" of contemporary capitalism, scarcely gives workers time to get back on their feet, let alone fight. Decelerationism is not a withdrawal to a slower pace of life, but the manifestation of an antagonism toward the progress of elites at the expense of the rest of us. It is Walter Benjamin's emergency brake. It is a wrench in the gears. The argument for decelerationism is not based on lifestyle, or even ethics. It is based on politics.

One of the biggest challenges facing the weak and fragmented Left is how to compose itself as a class—how to organize diverse sectors of people to mobilize for fundamental social change. This is due to changes in the technical composition of capital that create new challenges for worker politics: the erosion of stable jobs; the use of digital technology to proliferate work tasks; the introduction of the precarious, on-demand economy; the reinvention of scientific management practices; the massive financial and ideological power of tech companies. Through Luddism, we can challenge some of these forces, and, as workers in the nineteenth century did, begin to discover our common goals—and our common enemies.

"Luddism is not only popular; it also might just work."

In this way, Luddism is not simply opposition to technological innovation, but a set of concrete politics with a positive content. Luddism, inspired as it is by workers' struggles at the point of production, emphasizes autonomy: the freedom of conduct, the ability to set standards, and the improvement of working conditions. For the Luddites specifically, new machines were an immediate threat, and so Luddism contains a critical perspective on technology that pays particular attention to technology's relationship to the labor process. In other words, it views technology not as neutral but as a site of struggle. Luddism rejects production for production's sake. It is critical of "efficiency" as an end goal, as there are other values at stake in work. Luddism can generalize; it is not an individual moral stance, but a series of practices that can proliferate and build through collective action. Finally, Luddism is antagonistic. It sets itself against existing capitalist social relations, which can only end through struggle, not through factors like state reforms, the increasing superfluity of goods, or a better planned economy.

Ruptural Unities

Currently people are practically unanimous—they want to decelerate. A Pew Research Center poll found that 85 percent of Americans favored the restriction of automation to only the most dangerous forms of work. Majorities oppose algorithmic automation of judgement in parole cases, job applications, and financial assessment, even when they acknowledge that such technologies might be effective.

In spite of pop accelerationist efforts to re-enchant us with technological progress, we do not live in techno-optimistic times. Luddism is not only popular; it also might just work. Carl Benedikt Frey, the economist who sparked panic with his claim that 47 percent of jobs would evaporate by 2034, has recently acknowledged the Luddite wave. "There is nothing to ensure that technology will always be allowed to progress uninterrupted," Frey writes in *The Technology Trap*. "It is perfectly possible for automation to become a political target." He notes a variety of Luddite policies from the Left: Jeremy Corbyn's proposed robot tax in the United Kingdom; Moon Jae-in's reduction of tax incentives for robotics in South Korea; and even France's "biblio-diversity" law, which forbids free shipping on discounted books, to better preserve bookstores from competition with Amazon. History is full of such reforms against the worst tendencies of technological development, and they will be an important component of the coming deceleration.

A number of significant Luddish developments have been unfolding in recent years. One of the most promising is the surge in militant organizing within Silicon Valley against harmful technologies and for the rights of blue-collar tech workers. Beyond the tech industry, Luddite politics could link up with a number of emerging critical intellectual and political struggles, especially movements to address the environmental crisis. Green Luddism could be an alternative to the dead ends of technological solutionism and back-to-nature primitivism: a search for slower, less intensive, less estranged, more social methods of meeting our needs. Luddism might also link with the politics of degrowth, a movement that originated in the Global South and shares with Luddism an acknowledgment that liberation is not tied up with the endless accumulation of capital, and, further, that well-being cannot be reduced to economic statistics. Other contemporary points of resonance with decelerationism include the Maintainers, a research network that seeks to shift the focus of technological discourse away from "innovation," toward the vital practices of care and repair of existing technological infrastructures. Likewise, the "right to repair" movement, a Luddish technological initiative that advocates the conservation-minded maintenance of all sorts of digital technologies, from laptops to computerized farm equipment.

> *"Effective radical politics doesn't follow from an airtight plan constructed ahead of time with a specific revolutionary subject in mind."*

To be sure, these contemporary projects are vibrant, diverse, and, in some sense, incommensurate with one another. The same is true of many historical Luddish movements. Luddism manifests itself differently according to context. It is not a political program that various organizations and initiatives have signed on to in advance, but something more inchoate, a kind of diffuse sensibility that nevertheless constitutes a significant antagonism to the way that capitalism operates. And it can precipitate into concrete coalitions in unexpected ways.

Effective radical politics doesn't follow an airtight plan, constructed ahead of time with a specific revolutionary subject in mind. Even victorious revolutions are haphazard things, where disparate antagonisms build up, merge, and fragment. Louis Althusser, studying Lenin's analysis of the success of the Bolshevik Revolution, argued that it was not a case where the proletariat simply became sufficiently large and organized to overthrow the state. Rather, the revolution was a "ruptural unity": "an accumulation of 'circumstances' and 'currents'" many of which would "necessarily be paradoxically foreign to the revolution in origin and sense, or even its 'direct opponents.'"

As the cultural theorist Stuart Hall put it in his own reading of Althusser,

> The aim of a theoretically-informed political practice must surely be to bring about or construct the articulation between social or economic forces and those forms of politics and ideology which might lead them in practice to intervene in history in a progressive way.

My hope is that recognizing Luddism at work—in the office, on the shop floor, at school, and in the street—aids the ambitions of contemporary radicals by giving anti-technology sentiment historical depth, theoretical sophistication, and political relevance. We may discover each other through our myriad antagonistic practices, connecting to other struggles against the concentrated power of capital and the state.

To do so requires no preconstructed plan, no litmus tests of what is necessary in order to be properly political, authentically radical, or legitimately Left. As Marx put it in a letter to the Dutch socialist Ferdinand Domela Nieuwenhuis in 1881, "The doctrinaire and necessarily fantastic anticipations of the program of action for a revolution of the future only divert us from the struggle of the present." Rather, the first step of organizing disparate grievances into a collective politics requires recognizing and recovering our own radical self-activity, along with that of others. ∿

Gavin Mueller is a lecturer in New Media and Digital Culture at the University of Amsterdam.

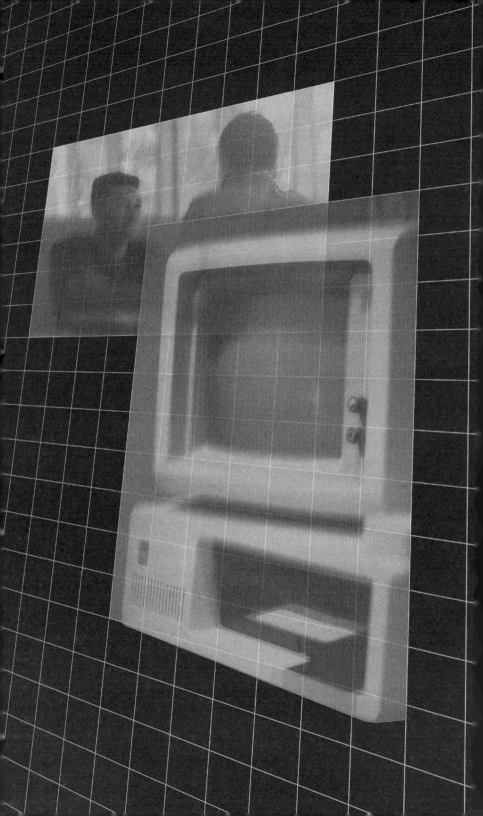

The Fort Rodman Experiment

by Charlton McIlwain

In 1965, IBM launched the most ambitious attempt ever to diversify a tech company. The industry still needs to learn the lessons of that failure.

————

Breonna Taylor, Ahmaud Arbery, and George Floyd were murdered in early 2020, victims of persistent anti-Black violence. In the midst of nationwide uprisings over their deaths, leaders in the technology industry responded. Amazon donated $10 million—roughly forty-five minutes worth of its gross annual profits—to racial justice organizations such as the National Association for the Advancement of Colored People. Social media companies like Facebook and streaming services like Netflix made content created by Black people more visible. Sundar Pichai, the CEO of Google and its parent company, Alphabet, promised a number of corporate commitments to racial equity, such as establishing anti-racist educational programs within the organization. And almost all of these actions were accompanied by pledges to bring more Black and brown people into tech company ranks—a desperately needed measure in a chronically white industry.

When tech leaders made those pledges, they often presented themselves as breaking bravely with the past: they would take unprecedented steps to overcome the implicit bias within their own companies and the structural racism of the industry as a whole in order to forge a more equitable future. ("Google commits to translating the energy of this moment into lasting, meaningful change," Pichai wrote in a letter to the company.) But there's good reason to doubt this self-presentation. In 2014, following pressure from public figures including Reverend Jesse Jackson, Google, Facebook, Apple, and Microsoft publicly disclosed their diversity data. Only 6 percent of Apple's workforce, 2 percent of Microsoft's, and 1 percent of Google's and Facebook's identified as Black, according to statistics compiled by *Wired*. Each company vowed to do better; Apple's Tim Cook said the company would become "as innovative in advancing diversity as we are in developing products." By 2018 and 2019, however, the percentage of Black tech workers at Facebook, Google, and Microsoft had increased by only one point; Apple's numbers hadn't changed at all.

Far from making a break with the past, when tech leaders pledge to diversify their companies, they are drawing from a playbook drafted over the course of the industry's history. Information technology firms have been trying—and largely failing—to become more racially representative since at least the 1960s. To understand some of the reasons why the tech industry has failed to become more diverse year after year, decade after decade, it's useful to go back to the earliest large-scale efforts by a major technology company—IBM—to diversify its workforce.

IBM has been actively trying to bring more Black and brown people into its workforce longer than any other major tech company, and it has adopted or invented the widest range of strategies to do so. If any company has had a margin of success in this, it's been IBM, and all of the tech companies that have come after it have in some way followed its example. At the same time, IBM's history is instructive because the company has been at the forefront of producing racist information technologies that have disparately harmed the very same people the company has spent decades trying to recruit—a dynamic that also characterizes many of today's tech giants.

> "When tech leaders pledge to diversify their companies, they are drawing from a playbook drafted over the course of the industry's history."

IBM's flawed motives, failed strategies, tempered successes, and massive contradictions over more than half a century provide critical lessons for today's tech industry to learn from if it is serious about advancing racial justice and equity. What the history of IBM shows is that creating racial equity in tech requires a commitment from institutions beyond the industry. It also demands that we rethink the sorts of technology that we allow tech companies to build.

The Original Bootcamp

In 1964, US civil servants transformed a former army base known as Fort Rodman, on the outskirts of New Bedford, Massachusetts, into the campus for an audacious new experiment in technical education. The base would host hundreds of male high-school dropouts—most Black, some white and

Latino, all poor or working-class—from across the country for a free fourteen-month training program designed to produce graduates who could go on to entry-level jobs at tech companies, including IBM. As IBM president Tom Watson Jr. later recalled in his memoirs, "The idea was to train 750 hard-core unemployed each year—black high school dropouts from the inner city who had never held jobs."

This was IBM's first and, in many ways, most ambitious diversity initiative. It was run by IBM but funded by the federal government as part of the Job Corps, a free education and workforce training program that was conceived by the Kennedy administration and which later became a key part of President Lyndon Johnson's Great Society, including the so-called War on Poverty. The Job Corps was one of the ways in which Johnson sought to quell the social and economic frustration fueling Black and working-class political mobilization during the civil rights era in order to maintain and expand Democratic power.

> *"As IBM president Tom Watson Jr. later recalled in his memoirs, 'The idea was to train 750 hard-core unemployed each year.'"*

IBM had two major goals in launching the Fort Rodman experiment. First, it hoped to ingratiate itself with the federal government, a source of lucrative contracts for everything from tabulation machines for the US Census to computer consoles for operating NASA space flights. Second, and perhaps more critically, the computing giant needed to train a large entry-level technical labor force to help fuel the company's rapid expansion. In 1965 alone, IBM acquired twenty-six new facilities, and in the subsequent five years it would double its annual revenue to roughly $7 billion ($45 billion in 2020 dollars).

The first crop of 350 students arrived at Fort Rodman in January 1965. They hailed "from the big cities and the small ones, the shut-down mining towns and the farm country" in New York, Texas, Alabama, and thirty-one other states, according to a 1966 promotional film about the program. Some of these young men may have been lured to Fort Rodman by postcards that featured an aerial photograph of the base on a sunny day, looking almost like a beach resort, and on the other side the text: "A HANDUP—NOT A HAND OUT." One student said that the program was his last resort; a judge told him it was either Fort Rodman, or else.

Students at Fort Rodman were separated into small cohorts, with one instructor assigned to five or six students. The instructors were white college graduates, some from the Peace Corps, who had been trained on site to be "tutor-counselors" to the young men who for more than a year would make Fort Rodman their home. The tutor-counselors were expected to be mentors and to bond closely with the boys; they ate with the students, hung out in barrack-style dormitories where as many as fifty slept in bunk beds with military-cornered sheets, and played football together. This was as much a part of the students' training as their remedial math and language courses and their regimen of office skills training, which included how to use typewriters, calculators, and keyhole-punch and data-processing machines.

In the 1966 promotional film for Fort Rodman, students seem impressed as an IBM employee shows them around a new punch-card tabulating machine. "How much time would this machine save compared to how you do 'em by hand?" a Black student in a shirt and blazer asks in the film. "Take a payroll application, for example," the IBM instructor answers. "A payroll that might take an entire week to prepare could be done on this machine in, say, two to three hours at the most." But if the electronic magic showcased by IBM captivated the boys at Fort Rodman, it wasn't enough to help them develop proficiency in the skills needed to get a technical job at a company like IBM. Most of what we know about the program comes from promotional materials that reflect how the people running the program, and IBM leadership, idealistically imagined it working. Even in these sources, however, the causes of Fort Rodman's failures are clear. Some were operational: for example, despite being designed to provide small-group, individualized attention, the program hired too few staff to meet student needs. Reports noted that students were often neglected by their instructors. Some students stopped showing up for class.

Compounding this neglect was no doubt the paternalism of Fort Rodman's mission, and the belief in Black cultural inferiority that the project embodied. In the 1966 promotional film, for example, as the camera fixates on the face of a young Black man working through a math lesson, the narrator intones:

> No one has ever given a damn about him until now. He's failed in school. He's failed with his family. He's failed within society. And so he is turned inwards and in a very bad way. We have to convert this history of serious failure into a present history of success.

This viewpoint—that it was the young men at Fort Rodman who were broken and needed fixing, not the systems of racist and sexist capitalism that Fort Rodman was, in theory, training them for—reflected the "culture of poverty" idea underlying many of President Johnson's Great Society programs. This idea held that Blacks were poor because they had an inferior culture that didn't prioritize work and individual responsibility, among other things; in order to change, Black people had to experience and adopt the "right"—supposedly white—cultural values. At the same time, Fort Rodman isolated young men from the communities that provided acceptance, care, safety, and pride for who they were as people.

For its part, the local community in New Bedford made it clear that the young men weren't wanted there; in May 1966, worried about "unruly elements" at the camp, as a *Washington Post* report put it, the city council asked President Johnson to move the Job Corps center out of Fort Rodman. Though Fort Rodman had enrolled more than 870 young men by then, the Johnson administration pressured IBM to close it. "The experience caused us some real soul-searching, because there were more problems than we anticipated," IBM President Tom Watson admitted in his memoirs. "IBM ended up hiring very few Camp Rodman 'graduates,' and I doubt any other company did either."

Racism as a Business Model

Fort Rodman may have been a failure, but IBM invented a number of other diversity programs that continued, with limited success, into the late 1970s. Several of its initiatives were aimed at luring Black people to IBM through job fairs and targeted advertising in Black media outlets, as well as by loaning equipment and funding faculty positions at historically Black colleges and universities. The company's primary focus, though, was on developing the "supply side" of the labor market by training the folks it hoped would fill its demand for technical workers. These efforts were smaller in scale than Fort Rodman, but similar in spirit.

IBM doesn't seem to have tracked its diversity programs with any rigor, making it difficult to know just how many they ran, where, and to what effect. But between 1978 and 1981, the period when IBM was most public about the success of its diversity programs, roughly 20 percent of IBM's new hires were non-white, and the number of non-white managers in the company increased from 1,973 to 2,600.

But like Fort Rodman, IBM's other diversity programs were flawed in important ways. Most notably, they focused on short-term, skills-based training for people whose educational background made it unlikely they would move beyond low-level positions at the company. Frank Cary, the chairman and CEO of IBM throughout most of the 1970s, admitted as much in a speech to the company's board and stockholders in 1974. "We've made good progress on one of our objectives—bringing into IBM capable and highly motivated minorities and women," Cary said. "Our second objective is taking longer to achieve: helping minorities and women qualify themselves for advancement at every level of the business consistent with their abilities and their growing population in the company."

Among the obstacles to promoting talented women and people of color, Cary's comments implied, was the desire among members of IBM's managerial class to hold onto the privileges conferred by their whiteness. "The relevant question I'm asked most frequently by IBM managers," Cary said, "is: 'How can we do that without practicing reverse discrimination?'"

This attitude was connected to a more fundamental problem. As much as IBM did in this period to try to remove the meta-phorical "Whites Only" sign from its company doors, racism at the company wasn't just a cultural or structural issue—it was part of its long-term business model. As early as the 1920s, IBM marshalled its computing powers to support eugenics, sterilization, and population control in Jamaica. The company sold technology to Hitler's regime that allowed the Nazis to tabulate census figures in order to identify and eventually murder Jews, and it sold similar technologies to South Africa to run the apartheid state.

> *"Racism at the company wasn't just a cultural or structural issue—it was part of its long-term business model."*

From about 1961 through the late 1960s, IBM was also deeply invested in helping federal, state, and local governments imagine, develop, and deploy carceral technologies that became known as "criminal justice information systems." IBM engineers, designers, and salesmen aggressively marketed computer hardware and software applications to the law enforcement community. Through lucrative contracts with big city police forces like the NYPD, research and development partnerships through President Lyndon Johnson's 1965 Crime Commission, and millions of dollars in grants from the newly formed federal Law Enforcement Assistance Administration, IBM laid the foundations on which today's policing and surveillance infrastructure has been progressively built over the past fifty years.

In 1968, for example, IBM debuted a system called ALERT II in Kansas City, Missouri. The system began as a database—a place to store police records about arrests, adjudications, jailings, and juvenile justice cases. But by the early 1970s, when it was fully built out, ALERT II, along with similar systems across the United States, was a nationally networked platform that provided law enforcement the ability to profile, surveil, target, and deploy police manpower based on the racial composition of neighborhoods and locations where crime allegedly predominated.

This reinforced a vicious cycle of racist policing. Because police believed Black people committed more crime, they deployed more police to Black neighborhoods. That led to more arrests, which meant Black people were captured more in police databases. Relying on that data to determine where to target police resources meant policing Black neighborhoods more intensively, thus perpetuating the cycle. As a result, entire communities were effectively criminalized in part by the technologies IBM was building.

"There is a clear thread connecting IBM's diversity projects with the racist technologies it developed."

Over the next three decades, through the 1970s, 1980s, and 1990s, such criminal justice information systems—some built by other companies following IBM's lead, many built by IBM itself—proliferated throughout the US, criminalizing Black and brown communities across the country. Since then, IBM has developed newer technologies with even more expansive law enforcement applications, including facial recognition, predictive policing, and police management systems—all of which wreak havoc on Black and brown people in similar ways to IBM's earlier generation of carceral technologies.

It's impossible to draw clear lines of causality between the racial makeup of IBM and the racist carceral technologies it has built. Would an organization with more Black and brown people in roles with seniority and power necessarily eschew helping law enforcement agencies criminalize communities of color? Would an organization that didn't build racist carceral technologies have more Black and brown people eager to join its senior ranks? Or would class interests trump racial solidarity so that even an IBM that was more diverse at all levels would still choose profit over racial justice?

Those may be unanswerable questions, but there is nevertheless a clear thread connecting IBM's diversity projects with the racist technologies it developed. In both cases, IBM saw Black and brown people as easily exploitable sources of profit—either in the form of low-wage labor, or as the material inputs that fed its policing technologies.

Building Black Tech

In the half century since the Fort Rodman experiment ended, big tech companies have launched many other diversity programs. But the numbers of Black and brown people in those companies, and the underlying logics of racialized capitalism that powers the technology industry, have remained largely unchanged. IBM's supply-side labor programs continue in the form of legions of coding bootcamps that promise Black and brown young people entree into the tech industry—though, in the absence of government and philanthropic support, these are run almost entirely as for-profit ventures.

The same approaches have correlated with the same results. The percentage of full-time Black employees in the tech industry today is about the same as IBM's was in 1965—roughly 2.5

percent. This lack of progress is reflected in, and may in part be caused by, the attitudes of the people who run these companies: a new report, People of Color in Tech, reveals that the majority of tech founders and CEOs believe that diversity work is ineffective. Roughly half of that same group are unconcerned about the fact that only 1 percent of tech entrepreneurs funded by VCs are Black. This sort of indifference is echoed in the experience of Black tech workers, who (more than their white peers) say they have trouble finding mentors at the companies they work for.

What would a more effective approach to improving the diversity of the tech industry look like? Three lessons stand out from the history of IBM's diversity programs. First, we need to ditch the supply-side approach that only prepares people for the lowest-level jobs, with the goal of creating an expendable and increasingly cheap labor force. Second, we can't leave it to the tech industry to change itself—we need government watchdog agencies like the Equal Employment Opportunity Commission to hold companies accountable, and the commitment of public resources, like those marshaled by Johnson's Great Society programs, to help transform society itself. Finally, we can't just seek to change racist cultures or structures at tech companies—we need to fundamentally change their business models.

In June 2020, IBM's recently appointed president, Arvind Krishna, announced to Congress that the company would no longer sell or develop facial recognition technology. He did so out of an explicit concern for racial justice, recognizing that these technologies have been and continue to be used to devastate people of color, including those within his own company. The tech industry should follow IBM's lead in examining its products, investments, and research and development projects. When these are inconsistent with racial equity and justice, the companies must abandon them. Diversity in tech is not just about sharing the gains of technology. It's about reimagining the tech we build and why. ⌇

Charlton D. McIlwain is a professor of Media, Culture and Communication at New York University, founder of the Center for Critical Race and Digital Studies, and the author of *Black Software: The Internet and Racial Justice, from the AfroNet to Black Lives Matter.*

Inside the Whale

An Interview with an Anonymous Amazonian

Amazon is on a mission to own the infrastructure of our lives. The second-largest private employer in the United States after Walmart, the company captures $4 out of every $10 spent online. They have a vast network of fulfillment centers, and they are rapidly buying more real estate; in September 2020, they announced plans to open 1,500 more distribution hubs in suburbs across the country. Retail is only one of their many businesses, however: for many Americans, it would be impossible to commute from home to school or the office without passing into view of a Ring, Amazon's "smart" home surveillance camera. Moreover, Amazon Web Services (AWS) controls nearly half of the public cloud market, and the company is pouring money into a number of other ventures, from entertainment to advertising. We talked with an AWS cybersecurity engineer about how to think about the behemoth and how it feels to work inside it.

———————

Amazon is a huge and complex organization. How should we think about it as a whole?

Amazon is an opportunistic corporation. It invests in businesses where we think we have a competitive advantage. In general, Amazon thinks of itself as a technology company. So we put the technology first, whatever the product is that we're selling. And we believe that because we have so much talent and so much capital, we should be able to use our technology advantage to dominate any market that we decide to enter.

What were its origins? Why did Amazon start off as a company that sold books on the internet?

In the mid-1990s, the internet was widely seen as a replacement for the library—the library 2.0—so figuring out how to buy books on the internet felt like a natural next step. A little later on, you could see some of that same spirit living at Google through the Google Books project, which was an enormous undertaking. They put a hugely disproportionate amount of resources into it. Amazon's ultimate goal was similar to that of Google Books: to digitize all of the information in the world's books and make it available universally, because that was the promise of the internet.

Jeff Bezos studies other "great men" in history and imagines himself to be a kind of Alexander the Great. There's even a building on the Amazon campus called Alexandria, which was the name of one of the company's early projects to get every single book that had ever been published to be listed on Amazon.

But there was also a more practical reason. Books are ideal because you can stuff them in a box. They're relatively cheap to ship. Also, they're easy to protect when shipping. It's difficult to damage books.

From the beginning, Amazon sold physical things. That meant its business evolved very differently than that of Google

or Facebook, which make their money by tracking people around the internet and using that information to sell ads.

Right, Amazon is not primarily an ad-driven platform. It does have a subsidiary, A9, that's in the online advertising business, but A9 is not a top moneymaker. The top money makers for Amazon by revenue are the retail side, and AWS.

> *"Jeff Bezos studies other 'great men' in history and imagines himself to be a kind of Alexander the Great."*

Why did Amazon get into the cloud computing business in the first place? What was the original impetus?

Amazon wanted to explore the possibility of selling web services because they realized most other firms weren't doing a terribly good job of it. From the start, startups flocked to AWS because we saved them a lot of time and effort. Once AWS had the start-ups hooked, it was easy to start selling to large businesses—the "enterprise" market—because they envied how well the more technically sophisticated startups were doing.

That was good for us, because big companies are more lucrative. But they also have stricter security requirements. They tend to be in mature industries that are more heavily regulated, and regulators care about how they're securing their data.

And that's where you come in.

For a long time, security hadn't been a big focus at Amazon because the data being collected—what books people were

ordering—wasn't that sensitive. It's not information most people were concerned about anybody having. We had to have a way
to secure credit card information to make online transactions
possible. But we outsourced that.

After AWS got started in 2006, security became a much bigger
concern. Amazon realized how important it was to its most
lucrative customers. These days, the company takes security
extremely seriously. I think you'd be hard-pressed to find many
nation states that have as sophisticated a security approach as
Amazon does.

Security is mostly about making yourself a difficult target. It's
like that joke where you go hiking with your friends and a bear
attacks you. You don't need to be faster than the bear; you just
need to be faster than your slowest friend.

**Big companies have traditionally operated their own data
centers. Was it hard to make the case to them that they
should move to the cloud? They might feel more secure if
they're doing everything themselves.**

They might, but ultimately the security standards of their data
centers are always going to be lower than those of a cloud
provider like AWS. A cloud provider has many tenants and they
can have economies of scale that let them have more sophisticated security systems than someone fully managing their
systems in-house.

Also, if you're a company that's operating your own data center,
you're responsible for 100 percent of your security—infrastructure security, transit security, perimeter security, everything.
If you move to the cloud, Amazon is responsible for at least
some of that.

So when you use AWS, part of what you're paying for is security.

Right; it's part of what we sell. Let's say a prospective customer comes to AWS. They say, "I like pay-as-you-go pricing. Tell me more about that." We say, "Okay, here's how much you can use at peak capacity. Here are the savings we can see in your case."

Then the company says, "How do I know that I'm secure on AWS?" And this is where the heat turns up. This is where we get them. We say, "Well, let's take a look at what you're doing right now and see if we can offer a comparable level of security." So they tell us about the setup of their data centers.

"You'd be hard-pressed to find many nation states that have as sophisticated a security approach as Amazon does."

We say, "Oh my! It seems like we have level *five* security and your data center has level *three* security. Are you really comfortable staying where you are?" The customer figures, not only am I going to save money by going with AWS, I also just became aware that I'm not nearly as secure as I thought.

Plus, we make it easy to migrate and difficult to leave. If you have a ton of data in your data center and you want to move it to AWS but you don't want to send it over the internet, we'll send an eighteen-wheeler to you filled with hard drives, plug it into your data center with a fiber optic cable, and then drive it across the country to us after loading it up with your data.

What? How do you do that?

We have a product called Snowmobile. It's a gas-guzzling truck.
There are no public pictures of the inside, but it's pretty cool.
It's like a modular datacenter on wheels. And customers rightly
expect that if they load a truck with all their data, they want
security for that truck. So there's an armed guard in it at all
times.

It's a pretty easy sell. If a customer looks at that option, they
say, yeah, of course I want the giant truck and the guy with a
gun to move my data, not some crappy system that I develop
on my own.

Wow.

There are also specific security services that AWS sells, such as
Amazon Inspector. Amazon Inspector is a tool that will audit all
of your configurations for AWS, and will provide recommendations about how to change those configurations.

When you make a connection to a server, that connection is
made over a specific port. And there are some ports that nefar-
ious people might sniff to see if they've been left open because
they're frequently used for the management of that server. And
so Amazon Inspector might say to you, "We have scanned your
server and detected that this port has been left open and you're
not using it. Do you want to close this port to prevent people
from trying to connect to it?" Or it might say, "These two servers
that you have are communicating to each other in a format that
is easily eavesdroppable. We recommend that you use at least
version X of the connection software that will plug some security holes."

If you're a competent system administrator, you should have
done all that when you configured your system. But not every

system administrator is competent. If you're the sysadmin for some, you know, insurance agency, what do you give a shit? You live in Sioux Falls. Why would you care about cloud security? You don't have to bust your ass because you live in a small-town market where you know everybody and you're never going to be out of a job. A lot of companies that are headquartered in remote areas don't have particularly sophisticated IT teams. So they'll pay Amazon to do security for them.

You mentioned that the most lucrative customers for AWS are large companies in mature industries, which tend to be more heavily regulated. How does AWS help those kinds of companies meet their compliance obligations?

Certain institutions and industries are regulated more than others. Take the healthcare industry. Hospitals and health insurance providers are bound by the Health Insurance Portability and Accountability Act (HIPAA). There are a lot of regulations about how data can be stored, how long it can be stored, and what types of consent are required. It can be very difficult to implement a system that is fully compliant with HIPAA, but AWS has products that can help with that.

> ## "*If you're the sysadmin for some, you know, insurance agency, what do you give a shit?*"

Same thing with the General Data Protection Regulation (GDPR), the European privacy law. It requires organizations to handle personal data in very specific ways. One of the new rights that it creates is the "right to be forgotten." So if you're a company that does business in Europe, you're going to need a way to fulfill requests from customers who want their data deleted. AWS can produce a report that says, "Yes, that data was deleted; we can no longer access it."

So compliance can create lucrative business opportunities for AWS. Presumably those opportunities will grow, with new privacy regulations like the California Consumer Privacy Act, and more regulations expected from the European Union around data and AI.

AWS is a money machine, basically. Today, retail is something like 70 percent of Amazon's revenue. But AWS is 70 percent of the company's operating profits.

One of the reasons why the antitrust people are looking at Amazon is because Amazon is using highly profitable businesses where it has a really durable advantage in order to subsidize losses in other divisions that it uses to capture market share. Without an organ similar to AWS, a competitor like Walmart has to lower prices below the level of profitability to remain competitive. And they can only sustain those losses for so long.

What's an example of a division that AWS subsidizes particularly heavily?

Prime Video, for one. Jeff loves Prime Video because it gives him access to the social scene in LA and New York. He's newly divorced and the richest man in the world. Prime Video is a loss leader for Jeff's sex life.

A Really Big Deal

We're now many months into the COVID-19 pandemic. How has work changed for you?

The first thing everybody noticed was conferences getting canceled, and everybody was like, "Wow, I guess Covid is a really big deal!" Then we emptied out all the buildings. There was a bifurcation of the people who work in corporate and the people who work in the fulfillment centers. The white-collar folks are fine because we can just work from home. But there was a huge internal drive to make fulfillment centers safe.

> ## *"AWS is a money machine, basically."*

What was the outcome of those conversations? I know there have been a number of collective actions among Amazon warehouse workers around the issue of safety during the pandemic.

At the beginning of the pandemic, management brought in some consultants and scientists to analyze how disease is spread. We had one week to figure out what we were going to do and then three weeks to execute. Now there are temperature reading cameras. You can't go into Amazon buildings if your temperature is elevated. They figured out how to reduce touch as much as possible. They implemented ways to scan badges at a distance and ways to use barcodes instead of near-field communication readers. They also created a testing regimen.

There was initially no way to get tests but because of our industrial might and muscle, we were able to identify a vendor, so we worked with them to do tests for warehouse workers. I'm not sure how often those tests occur. And it's not like we test for a whole slate of things. A lot of the warehouse workers are older people who are out of the traditional workforce and find it hard to get back in because they can't retrain or they don't want to, or the job they used to do doesn't really exist anymore. A lot of them don't work enough hours to get health insurance. So if you have cancer and you might die from your cancer, we won't help you get treatment. But if you have this infectious respiratory disease, we want to know.

I actually don't know if the steps that have been taken around Covid have made things worse, or whether they have improved the warehouses. Internally, people say, "Oh, we're probably better than our competitions, or other warehousing and logistics companies." But I don't know if that's the case.

There were other challenges at the beginning of the pandemic too, right—with supply chains for instance?

Supply chains all across Amazon were definitely impacted. It was difficult to get hand sanitizer. It was difficult to get cardboard boxes.

We got lucky in the sense that the beginning of the pandemic overlapped with the Chinese New Year. So we had already accounted for some slowdown, because we expected the Chinese New Year to impact timetables anyway.

Overall, though, it seems like the pandemic compounded all of Amazon's advantages and significantly reduced the impact of all of Amazon's weaknesses.

How so?

The crisis advantaged large purchasers. The more scale you have, the more buying power you have, right? If you're a factory, the way that your distribution of customers is set up is that you have one or two customers that make up 80 percent of your business and then a ton of customers who make up 20 percent of your business. So if your production is down 30 or 40 percent, you say, "Okay, I'm going to have to cut out all of the small vendors and only focus on the big partners." That's what happened. Amazon was able to get preferential distribution of shipments when smaller companies were totally sold out.

And there's been so much e-commerce growth in general during the pandemic.

The question on everybody's minds in retail is: can Walmart and Target use their local distribution infrastructure to get packages to people's doors faster than Amazon can? Walmart and Target's only advantage is their physical stores. When the stores are closed, they can be used as distribution hubs. You can pick things up close to where people's homes are and deliver them. Buying Whole Foods gave Amazon the opportunity to reach, you know, 80 percent of the 1 percent. But vast swaths of America aren't reachable by Whole Foods.

The rumors that I hear, both internal and external, are that we're very seriously interested in acquiring post office real estate. The reason why the post office is valuable to privatize is because of their real estate holdings. They have great real estate in every downtown of every city in the United States. Amazon may be interested in buying all of the post office locations, and we have the cash to do it. So why not?

"The pandemic compounded all of Amazon's advantages and significantly reduced the impact of all of Amazon's weaknesses."

The other week we announced we're hiring one hundred thousand more workers again. We're expanding dramatically across the board, in part-time and full-time, at corporate and retail and fulfillment and logistics and devices and distribution and all the various pies we have our fingers in.

How about AWS? Are you growing the same way?

Some of our customers obviously saw a dramatic decline in their income. Some large customers in the hospitality industry and the retail industry negotiated substantially discounted rates or got large one-time credits.

But in general, COVID did convince a lot of companies to accelerate their cloud migration, because if you're an organization that has your own data center, chances are that you now have to implement a bunch of safety guidelines and restrictions. Amazon just operates at such a scale that we can do it better and more efficiently.

> *"In the end, the market is learning that massive data exposures are not that bad of a problem unless Congress comes calling."*

For instance, a lot of security companies throughout the industry were impacted because suddenly their secure facility had entry rules. Only one person could be in there at a time. But in the security technology industry there are a lot of processes that need two or more people to be physically present. So that made it impossible to use a lot of the secure facilities because the guidelines were in conflict. It was kind of a Freaky Friday moment where everybody realized that, in the context of a pandemic or a natural disaster, these procedures they created to make sure that their facilities were safe were actually preventing them from following security best practices.

Customers who were experimenting with a small presence in AWS, who had maybe kept their own data center for security purposes, freaked out. They were like, "Wow, we really need to move to the cloud quickly because they can do all these things for us that we can't do on premises now."

Besides the challenges of dealing with social distancing, are you seeing new cybersecurity threats in the Covid era?

The security threats that are emerging now are the same as ever. They're just more intense. 80 percent of security problems are petty cybercrime. And that's what's going up.

Why?

Increasingly, large tech firms use people in developing countries as a disposable white-collar workforce. Smaller shops do too. A lot of startups will have one CTO in the Bay Area, and then they'll have their whole development shop be in Ukraine or Romania or something.

But when funding dries up for startups and companies have to shutter, then all of their digital operation overseas is cut loose. And the people who lose their jobs go into cybercrime. They think, "There's no other options for me. So sure. Let's do it. Lock and load."

A cybercriminal can also get paid by a competitor to expose data, or to change the configuration so the data is exposed publicly to make that company look bad. The Capital One breach in 2019 involved a former Amazon employee, actually. Capital One suffered a huge embarrassment in the press. But unfortunately, as with the Experian data breach in 2015, the Capital One incident showed that the markets are very forgiving of data breaches, because the people who are most victimized by them are poor people who have no idea how to control their data anyway, and didn't even know what it means to have their data breached.

Companies don't like to have their whole ass be shown that way. It's a lot of egg to get on your face. But in the end, the market is learning that massive data exposures are not that bad of a problem unless Congress comes calling.

Unlocking the Last Foot

So you work in cybersecurity and that's clearly a major focus for AWS. But as we talk about this, there's another sense of the word "security" in the back of my mind—home security. This is an area Amazon is now getting into with Ring, the internet-connected doorbell.

I wasn't involved with that acquisition. But what I've heard is that our investment in Ring was initially about wanting to combat package theft.

The retail side of Amazon is basically a logistics company. We have a distribution infrastructure that we chop up into different segments. "First mile" is from manufacturing to distribution. "Middle mile" is from the first distribution center—a warehouse—to the second—the place where a package will get delivered from, like a postal depot. "Last mile" is from that second distribution center to people's front door. And one of the major problems with the last mile is package theft.

But that wasn't the only motivation with Ring. More broadly, Amazon's smart home projects are also aimed at unlocking the "last foot"—not just how you get the package to the customer's door but into their house. At one point we tried to make an electronic lock with an electronic key that Amazon deliveries could use. But then someone else made Ring, and we realized we could use that instead.

Amazon has aggressively marketed Ring to police, partnering with hundreds of law enforcement agencies across the country and in some cases even giving the devices away for free. Cops are given access to a portal they can use to request Ring footage from individual houses. Has Ring brought Amazon into much closer relationships with law enforcement?

Relationships with law enforcement take a very long time to build, so it would really surprise me if any of those relationships were the result of the Ring acquisition. The groundwork was already being laid. I think Ring just helped accelerate things.

In general, the nice thing about working with law enforcement is that they know what they want. Regulators don't.

But honestly, I think Amazon also kind of backed into that situation. We only realized after the fact that we had all this data about who was coming to people's front doors. And then there was a lot of gleeful Mr. Burns–style finger touching, when we thought about what we could do with that data. Ring has a Neighbors app, where you could take your Ring data and share it with the app, and your neighbors could see who was in the neighborhood. So law enforcement was a natural next step. And law enforcement dovetailed nicely with our interest in pursuing the home security angle through our other smart home products.

Like the Alexa-enabled smart speakers.

Right. Ring dovetails nicely with Alexa on the home security front. And Alexa could also help the retail effort to unlock the "last foot" and get packages inside the home.

You said that Amazon was already building relationships with law enforcement long before the Ring acquisition. Why?

AWS works extensively with US government agencies. At the federal level, AWS runs a special cloud for a number of intelligence agencies, and we're still trying to get the contract for JEDI, the Pentagon's big cloud project.

> "*It's the same twenty thousand people in the United States who have had jobs in the military, regulatory agencies, and industry, all selling these things to each other back and forth.*"

The other factor that presumably contributes to Amazon's coziness with intelligence agencies and law enforcement is the fact that many people from those fields go work for Amazon. Why is that?

Amazon pays a better salary and you get to work on more exciting stuff, and there's just less hassle.

Do the cultures mesh well?

In general, the military is pretty top-down, command-and-control. That's not Amazon. We mostly want people to be as autonomous as possible. At Amazon it's easy to shoot up the hierarchy and talk to senior executives if there's a problem.

Does Amazon make a concerted effort to hire from law enforcement and the military?

For the rank and file, yes. There is a concerted effort to recruit former law enforcement and military. In fact, Amazon thinks of military personnel as a diversity category and does targeted hiring. We have an internal affinity group called Warriors@Amazon for ex-military personnel, and it is by far our most successful diversity hiring group.

At the higher levels, there's a revolving door. If you're a chief procurement officer at the Pentagon, the guy who orders whatever they're buying for the military, you do that for awhile and then you go to the General Accountability Office to be a watchdog. And then after you've been a watchdog for awhile, you go to work for Amazon, where you can make half a million dollars a year selling Amazon services back to the Pentagon. And you can get it past the watchdog because you used to be the watchdog.

Amazon's not the only company that does this, obviously. Everybody does. It's the same twenty thousand people in the United States who have had jobs in the military, regulatory agencies, and industry, all selling these things to each other back and forth.

In September 2020, Amazon appointed Keith Alexander, the former NSA chief, to be on your board of directors. How did you and your coworkers see that?

At the most obvious level, it's a move to get influence over the Defense Department and win more US government contracts. Because Keith Alexander has relationships with all of the right people in the Pentagon, he can help move the needle in terms of not only managing relationships with current officials, but also helping us strategize about what's important to the decision makers going forward.

But it was also highly controversial. Even within Warriors@ Amazon, they're of two minds about it. Some of the Warriors love the US government. On the other hand, bringing on someone like Keith Alexander also poses serious risks to Amazon's business.

How so?

Our European partners are screaming about the message that hiring Keith Alexander sends with regards to the privacy of their data. It also comes at a particularly bad time. AWS needs the European market. But there have been a couple of recent legal decisions in Europe that have made our lives harder, like the courts striking down the Privacy Shield agreement in the summer of 2020. [*Eds.: Privacy Shield was a compliance framework that certified the data security of US-based companies, and allowed them to receive EU user data in a manner deemed compliant with GDPR.*] AWS has been working really hard with customers to make sure that we can comply with whatever new privacy standards the European Union develops.

So within this context, bringing Keith Alexander onto the board is definitely going to raise a lot of eyebrows on the other side of the pond. Our European partners are going to say, "Well, excuse me. Our concern is that you're taking data from our citizens and bringing it back to the States. And now you're hiring the architect of the program that's been spying on the entire world?!" It sends mixed signals at best. But the fact that there's a business case against it has also helped the people inside the company who are opposed to it for philosophical or ideological reasons.

Are there many of them?

A lot of my coworkers are concerned about Keith Alexander's presence.

I would say a quarter of the people at Amazon would identify with digital freedoms organizations like the Electronic Frontier Foundation, and they are pissed off. I think there are a lot of AWS employees who would rather quit than turn data over to the NSA. Those people are all up in arms and they're some of the smartest and most innovative people at the company. I don't think Amazon counted on so much internal opposition and resistance.

Inside Voices

White-collar workers at Amazon have organized internal campaigns against the sale of facial recognition software to law enforcement agencies, and the sale of cloud services to companies that enable ICE. Has Amazon's relatively large contingent of former law enforcement and military personnel made the organizing environment for these campaigns more challenging?

My experience with ex-military people is that they love America but they don't care for the government. So they're like, "Yeah, I worked for Uncle Sam, but I could give a shit." So I don't think the presence of those people necessarily makes organizing harder. I just think the collapse of the Left in general over the past several decades has made workers everywhere believe that resistance is futile. Maybe we're doing right in the world, maybe we're doing wrong—but if we're doing wrong, there's really very little we can do about it except some things around the margins.

It seems that most of the organizing energy at Amazon among white-collar workers has been around climate issues. There is an ongoing campaign to push the company to dramatically reduce its carbon emissions that have involved open letters to Bezos. And in April 2020, the company fired two white-collar workers who had helped lead the climate organizing, and who had also criticized the company's treatment of warehouse workers during the pandemic. What's your perspective?

I didn't sign any of those letters, because I'm not sure what good they accomplish either internally at Amazon or externally in the world.

In general, campaigns at Amazon fall into three groups. The first is mainstream causes that are palatable to everyone, like donating winter coats and protecting animal rights. The second is environmentalism. It's easy to find people at Amazon who care about global warming. So there is some support for people who recognize the severity of the problem and are willing to organize internally at Amazon around it. But that organizing effort is still pretty small. It has an outsized voice in the media, because everybody's fascinated by Amazon. But there aren't a ton of people who are paying attention to it internally.

"Amazon doesn't actively not want employees to talk to each other. They just don't see how employees talking to each other benefits productivity, morale, or the bottom line."

The third category of campaigns is social justice activism, like the #TechWontBuildIt campaigns against working with ICE or law enforcement. There's a spirited debate on a couple of internal mailing lists about that kind of activism. But again, it's a small number of Amazon employees.

Why do you think there aren't a lot of people participating in this conversation internally?

Historically, Amazon has probably had some of the worst internal communications tools of any large company. It's very difficult to discover active conversations happening in the company. There is no central clearinghouse. There is no internal social media. We recently got Slack, but it hasn't made much of a difference for employees interested in organizing.

Do you think the lack of internal communications tools is intentional?

I think this is one of those cases where you should not presume malice. Amazon doesn't actively *not* want employees to talk to each other. They just don't see how employees talking to each other benefits productivity, morale, or the bottom line. If it did, and that impact could be quantified in some way, we'd have it tomorrow. But as it is, Amazon gives employees the tools that it thinks will help them get the job done. And they don't see employee fraternization as relevant to the job.

Google, by contrast, has very robust internal communication infrastructure. And that infrastructure played an important role in facilitating organizing at Google. (Although more recently, management has been limiting the kinds of conversations that can happen on internal platforms.) Do you think the absence of similar tools helps explain why Amazon has seen comparatively less organizing?

Well, if you look at Google, you'll notice they're headquartered in Mountain View, in the heart of Silicon Valley. And if you're an employee at Google and you're good at your job and you want to leave your job tomorrow, there are fifty-three employers out there that are going to be ready to hire you.

If you're working in Seattle for Amazon and you're good at your job and you want to leave your job tomorrow, you have far fewer opportunities. Where are you going to go, *Microsoft*? There's not nearly as much mobility. So I think a big part of the reason we have less organizing is that people are more afraid to jeopardize their jobs. If you want to stay in the Northwest, you keep your head down.

Statistically it is also more likely that an Amazon employee will have a family than a Google employee. So that's another factor that makes people more risk-averse. Why should they do something that would potentially jeopardize their job? Particularly when it has a low chance of success?

As you pointed out, one of the reasons that the organizing efforts within Amazon have received so much media attention is because the media is fascinated by Amazon. There have been a spate of stories looking critically at Amazon's market power, partnerships with law enforcement, labor conditions in its warehouses, and so on. Amazon also has prominent critics in national politics like Bernie Sanders.

How are these kinds of criticisms perceived from the inside? How do people respond to that sort of thing?

I think your question kind of misses the forest for the trees. For most people at Amazon, glancing at the Apple News feed on their iPhone is about as much of the discourse as they consume. They don't care about the news. It doesn't contribute anything to their life. There are colleagues I'm friends with who don't really know who ran for president. They figure it's all going to be the same anyway, so why bother.

But by the same token, if they hear someone criticize Amazon, they're not inclined to be super defensive. There aren't a lot of intense loyalists. People at Amazon are mercenaries. The company doesn't have great benefits. Office life kind of sucks and it's not that fun of a place to work. It's a grind. People work there because it pays a little bit better than the competition and it looks good on a resume. They can go in, do their job, go home, spend time with their kids, watch sports. That's the good life.

> *"If you're working in Seattle for Amazon and you're good at your job and you want to leave your job tomorrow, you have far fewer opportunities. Where are you going to go,* **Microsoft?***"*

Amazon has around a million employees worldwide. The majority work in shipping and logistics and delivery. There are maybe eighty thousand corporate employees. And I would estimate that fewer than two thousand of them have participated in discussions around organizing.

Do you see any cause for hope?

In general, the people who are going to organize are the people who need to organize because they are fighting for their lives and their subsistence. Those are the people on the logistics side of Amazon who work at the distribution centers. Those are the members of the industrial proletariat in China who are manufacturing the things that are shipped out on the retail side. Those are the humans in developing countries doing piecework on Mechanical Turk.

If there is going to be change, that's where it will come from. I think that if you're looking at corporate employees within Amazon as a source of hope, that's ludicrous. The notion that these companies are going to repair the damage they're causing by having white-collar workers organize internally to me is crazy. But maybe that's cynical and nihilistic. Maybe I'm a bad man. ∿

How to Make a Pencil

by Aaron Benanav

Capitalism is over, if you want it.

———————

What would a socialist economy look like? The answers to this question vary, but most of them involve planning. A capitalist economy is organized through the interaction of prices and markets. A socialist economy, by contrast, would be "consciously regulated ... in accordance with a settled plan," to borrow a line from Marx. But how would such a plan be made and implemented? This has been a matter of sharp debate among socialists for more than a century.

One camp has placed particular emphasis on computers. These "digital socialists" see computers as the key to running a planned economy. Their focus is on algorithms: they want to design software that can take in information on consumer preferences and industrial production capacities—like a gigantic sieve feeding into a data grinder—and output the optimal allocations of resources.

Over the years, there have been a number of experiments along these lines. In the 1960s, the Soviet mathematician Victor Glushkov proposed a nationwide computer network to help planners allocate resources. With the help of the English cybernetician Stafford Beer, Salvador Allende's administration in Chile tried something similar in the 1970s, called Cybersyn. Neither project got very far. Glushkov's idea ran into resistance from the Soviet leadership, while Pinochet's coup ended Cybersyn before it was fully implemented. However, the dream lives on.

Today, digital socialism could obviously do much more. The internet would make it possible to funnel large quantities of information from all over the world into planning systems, almost instantaneously. Gigantic leaps in computer power would make it possible to process all of this data rapidly. Meanwhile, machine learning and other forms of artificial intelligence could sift through it, to discover emergent patterns and adjust resource allocations appropriately. In *The People's Republic of Walmart*, Leigh Phillips and Michal Rozworski argue that large companies like Walmart and Amazon already use these digital tools for internal planning—and that they now need only be adapted for socialist use.

While there are certainly emancipatory potentials here, they are far from adequate to the task of planning production in a post-capitalist world. The digital socialist focus on algorithms presents a serious problem. It risks constraining the decision-making processes of a future socialist society to focus narrowly on *optimization*: producing as much as possible using the fewest resources. To travel down this road is to ignore and discard vast amounts of qualitative information, which remains crucial to achieving many of the ends and goals of a socialist society.

After all, the societies of the future will want to do more than just produce as much as possible using the fewest resources. They will have other goals, which are more difficult to quantify, such as wanting to address issues of justice, fairness, work quality, and sustainability—and these are not just matters of optimization. This means that, no matter how powerful the planning algorithm, there will remain an irreducibly political dimension to planning decisions—for which the algorithm's calculations, no matter how clever, can only serve as a poor substitute. Algorithms are essential for any socialist planning project because they can help clarify the options among which we can choose. But human beings, not computers, must ultimately be the ones to make these choices. And they must make them together, according to agreed-upon procedures.

> "*Societies of the future will want to do more than just produce as much as possible, using the fewest resources.*"

This is where planning protocols come in. They streamline decision-making by clarifying the rules by which decisions are made. Deployed in concert with algorithms, protocols enable a range of considerations—besides those available to an optimization program—to enter into the planning process. We might say there is a division of labor between algorithms and protocols: the former discard irrelevant or duplicate options, clarifying the decisions to be made via the latter.

Putting both algorithms and protocols to work, people can plan production with computers in ways that allow their practical knowledge, as well as their values, ends, and aims, to become integral to production decisions. The result is something that neither capitalism nor Soviet socialism allowed: a truly human mode of production.

The Price Is Right

Any serious attempt at socialist planning has to reckon with the problems posed by the "socialist calculation debate," a decades-long argument that has influenced how generations of socialists have imagined a post-capitalist future. The right-wing Austrian economist Ludwig von Mises kicked off the debate in 1920 with "Economic Calculation in the Socialist Commonwealth," a full-frontal assault on the feasibility of socialist planning.

At the time, this wasn't just a theoretical question. The revolution was already well underway, not only in Russia, but also in Germany, and very nearly in Italy and other countries. Socialists claimed that, with the capitalists cast aside, they could use modern machinery to construct a new type of society, one oriented around human needs, rather than profit. Everybody would get access to the goods and services they needed, while working less.

Mises argued that socialists were wrong on both counts. Instead, people in a socialist society would work more hours and get less for it. That's because, in his view, the efficiency of modern economies was inextricably connected to their organization via the market, with its associated institutions of money and private property. Get rid of these institutions, and the technologies developed over the course of the capitalist era would become fundamentally worthless, forcing societies to regress to a less advanced technological state.

To illustrate Mises's point, let's take a simple example: the manufacture of a pencil. The manager of a pencil-making factory has to make many production decisions, because there are many ways to make a pencil out of its component parts. How does a pencil maker decide how to produce his "final good," the pencil, out of all the possible "intermediate goods," the various types of graphite, wood, paint, and other things that go into making it?

In a capitalist society, he begins by checking the price catalog, where he discovers that graphite A costs 35 cents per pound, while graphite B costs 37 cents. If either works, his choice is clear. This manager can perform the same price test for all the relevant inputs, in order to arrive, quickly and accurately, at the most rational way to make a pencil. He does not need to understand how all the activities of society add up to an overall economy.

Prices allow the pencil makers to quickly set aside numerous procedures for making pencils that would result in functioning pencils, but at the cost of squandering natural or labor resources better employed elsewhere. If given tons of the finest quality Cocobolo or Osage Orange lumber, the pencil makers could undoubtedly make good pencils. But this would be a waste if some other tree, like the humble cedar, provided lumber that worked just as well.

Of course, the prices that pencil makers use to make production decisions are not just random numbers. They are expressions of a living market society, characterized by decentralized decision-making, involving large numbers of producers and consumers. Markets place pressure on all producers to get prices right. If it proves possible, for example, to make pencils more cheaply without sacrificing quality by using a new technique, the firm that does so will earn a sizable profit. New information about pencil production possibilities will show up in the system as a lower pencil price.

Each producer can make rational decisions about what and how to produce, only because a struggle for market supremacy forces producers to maximize their revenues and minimize their costs. All of these market-dependent producers absorb information to the best of their abilities, make decisions, and take risks in search of new production possibilities and the corresponding monetary rewards. Socialist planners couldn't possibly reproduce such a complex system, Mises believed, because they would never have more information than market participants mediated through the price mechanism.

Ultimately, prices tell producers which production possibilities have any chance of turning a profit. Without prices, Mises argued, the rational allocation of assets becomes impossible.

Fatal Errors

What's striking about Mises' description of capitalism is that it is already highly algorithmic. In his account, the managers of the pencil factory behave like a computer program. They collect price information about intermediate inputs and then follow a simple rule: choose the cheapest option for each input that does not lengthen production time or lead to an unacceptable reduction in demand.

Many socialists responded to Mises's challenge by accepting his basic premise and then trying to write their own algorithm. In other words, they wanted to show that planners could create a substitute for the price system that could generate enough information to arrive at the correct production decisions for a socialist society.

The Polish economist Oskar Lange and the Russian-British economist Abba Lerner were the first to develop this idea. Their proposals, worked out over the course of the 1930s and 1940s, involved socialist planners "feeling" their way towards the right prices through trial and error. For example, planners might set the price of an intermediate good required to make a pencil, and then adjust that price as necessary, until the supply of the final good matched consumer demand. A series of approximations would get closer and closer to the true result, much like a computer calculating pi through a sequence of slight additions or subtractions.

"Without prices, Mises argued, the rational allocation of assets becomes impossible."

When Lange and Lerner were writing, modern digital computing didn't exist. But at the end of Lange's life, as computers emerged, he discussed the possibility that they could perform this price-guessing work far better than humans. This line of thinking has been taken up by contemporary digital socialists, who point to developments in applied mathematics as evidence that we could do away with the price system, calculating optimal allocations of resources with advanced forms of programming instead.

After all, we have more data than ever before, as well as an unprecedented amount of processing power with which to perform computations on that data. Gigantic firms like Walmart and Amazon are already using advanced algorithms to put all this data to work to plan their internal operations. So, can the promise of algorithmic socialism finally be fulfilled?

Not so fast. Advocates of algorithmic socialism misunderstand Mises's position in the socialist calculation debate, and thus fail to respond adequately to his criticisms. For Mises, the challenge is how to allocate intermediate goods to producers of final goods. That's not something companies like Walmart and Amazon do, for the simple reason that these companies distribute goods rather than make them. The firms supplying pencils to Amazon and Walmart still rely on market signals to figure out the best way to make their product.

As Mises's student Friedrich Hayek later emphasized, an economy is not a set of equations waiting to be solved, either with a capitalist price system or a socialist computer. It is better understood as a network of decision-makers, each with their own motivation, using information to make decisions, and generating information in turn. Even in a highly digitally mediated capitalist economy, those decisions are coordinated through market competition. For any alternative system to be viable, human beings still need to be directly involved in making production decisions, but coordinated in a different way.

As Hayek observed, running a business involves practical reasoning, acquired through years of experience. To reproduce the work of the manager of a pencil factory, a planning algorithm would have to know not only about the supply and demand for each type of graphite used in pencil making, but also about the detailed implications of choosing one type of graphite over another in that particular production location, with its specific machines and workforce. It is possible that one could formalize all of this knowledge into explicit rules that a computer could execute. However, the difficulties involved in articulating such rules across all workplaces, in all sectors, are simply staggering.

Mises and Hayek were correct to observe that people's participation in decision-making will remain essential for any economy to function. Yet their vision also sets strict limits on who has the opportunity to exercise this agency. In capitalism, the people involved in making production decisions are managers. They represent only a small fraction of the total number of people involved in production, and they do not need to consult all of those other people when making decisions—except insofar as they are forced to do so by law or contract.

> *"An economy is not a set of equations waiting to be solved, either with a capitalist price system or a socialist computer."*

Managers are therefore free to pursue economization within broadly defined limits. If their decisions require that large numbers of workers in a particular town lose their jobs—because the pencil factory is being moved to a place with lower labor costs, for instance—then that is a decision the manager can make without answering to the townspeople. For the market to function, therefore, decision-making power must be concentrated in relatively few hands.

In a socialist society, however, the entire population would control production. Decision-making power would be democratized, and this would almost certainly lead to different kinds of decisions being made. Should people begin to run their own workplaces, they would likely decide to introduce all sorts of changes, such as those related to working conditions, for instance, or to how tasks are organized and assigned. Efficiency, whether calculated in terms of energy use, resource consumption, or labor time, would remain a concern, but it would no longer be the sole concern. It would simply be one of many. Other considerations — dignity, justice, community, sustainability — would also enter the picture.

> *"Neurath argued that a socialist economy would have to be highly democratic—precisely because it could not be purely algorithmic."*

These other considerations could not easily be absorbed into a one-dimensional optimization algorithm, however, for the simple reason that there is no reliable way to reduce them all to a single, quantitative unit of account. Even natural units, such as tons of iron or grams of penicillin, would prove inadequate. What is the natural unit of justice? Given these constraints, the most advanced computer on the planet still could not determine the correct production plan because the different choices are rooted in competing values and visions of the good — in other words, they are political choices.

If socialist planning is purely algorithmic, it executes decisions in a similar way to capitalist firms. It reiterates the logics of capitalism in a different register: what matters is the extraction of the relevant quantitative information from the mess of qualitative life. But it is only in this mess that the content of socialism can be found.

Crafting the Protocol

How can a greater variety of qualitative goals become part of the planning process, to be pursued for their own sake? To answer this question, we need to turn to the work of Viennese philosopher Otto Neurath.

Neurath was one of the original targets of Mises's 1920 broadside against planning. He is remembered today as the theorist of total planning—a phrase that incorrectly conjures the image of social engineers running the economy from a control room. Nothing could be further from Neurath's vision. On the contrary, Neurath argued that a socialist economy would have to be highly democratic—precisely because it could not be purely algorithmic.

For Neurath, the algorithmic character of the price system was a problem to be overcome, rather than something that socialists should try to replicate. In a capitalist economy, managers are able to make clear-cut decisions about cost-effectiveness only because they are allowed to ignore all of the non-economic costs of their decisions, which include destroying communities, immiserating workers, depleting non-renewable resources, and filling the world with garbage. Economically rational decisions at the level of the firm add up to an increasingly irrational society.

Instead of just optimizing for efficiency, then, socialists need to figure out how to incorporate multiple qualitative criteria directly into their planning mechanism. The issue socialists face is not quantification as such. They could probably quantify many of the criteria relevant to their production process—establishing indexes of sustainability and safety, for example. But to distill all such relevant indicators to one unit of account suggests a degree of commensurability between goals that is exactly what socialists would want to overcome.

A capitalist society that wants to reduce pollution needs to set legal limits on how much each factory can pollute, allowing those firms to continue to optimize their production strategies, but under new restrictions. That, in turn, creates incentives for pencil factories to get around those restrictions—and if they can figure out how to pollute without getting caught, those firms can make large profits. By contrast, a socialist society would want to take pollution reduction as a goal to be pursued for its own sake. It would look for ways not just to limit pollution at the pencil factory but to positively improve the environment—increasing air quality, planting trees, and so on—wherever doing so does not rule out the pursuit of other goals.

Such an approach requires far more than mere optimization. Rather than trying to convert all of the qualities and quantities of life into a unifying metric that can be algorithmically optimized, we need to find a way to deal with those qualities and quantities on their own terms. We need to be able to make planning decisions on the basis of multiple, incommensurable criteria, and to coordinate these decisions across society. To do this, we must have agreed-upon procedures for making such decisions collectively—protocols.

There are many ways to design a planning protocol. It could be as simple as a population-wide vote, with the majority deciding the outcome. Or it could take the form of a complex bidding procedure, like an auction. A protocol could even be a game, with a set of rules that specifies who can play, what actions each player can take, and what real-life allocations result from different outcomes. There are many possibilities, but the unifying theme is the need to to craft protocols that allow actual human beings to make holistic decisions that take a variety of criteria into account.

Neurath laid out his version of a planning protocol—a term that he did not himself use—in "Economic Plan and Calculation in Kind," an essay he wrote in 1925. Planning begins with expert planners reducing the "unlimited number of economic plans" that are "possible" down to a few "characteristic examples." These planners do the algorithmic calculations, which clarify the options among which people must decide. People are then presented with these options for direct comparison. They evaluate a few different plans across multiple criteria and decide which they prefer: listening to comments, voicing their concerns, and taking a vote.

"*Such an approach requires far more than mere optimization.*"

Neurath believed that such a process would enable a particular kind of rationality to emerge. Even where it proves impossible to make clear and precise calculations, he argued, we can still decide rationally. However, the rationality we deploy will be a practical and political rather than purely algorithmic. People will have a chance to voice both their concerns and their desires, before arriving at collective decisions about how to shape, constrain, and direct the production process. They will balance how much they want to consume against how much they want to work. They will weigh their need for energy to heat their homes and power their workplaces against values of ecological sustainability and intergenerational justice. They will decide how much of their time and resources would be set aside for expanding or transforming production and how much for cultural, athletic, and intellectual activities.

"True democratic decision-making about production cannot simply be a matter of a perpetual social-media plebiscite scrolling across one's phone screen."

In Neurath's model, decisions made collectively, at the highest level, would then filter down through the rest of the economy, to be implemented across various industries and workplaces. But how would that work exactly? How are local production decisions made? What happens if conflicts or collisions arise—for instance, between the decisions of society as a whole and the demands of workers in pencil factories, producing goods to meet society's needs?

These complexities suggest that what we need is not one society-wide protocol but many protocols—many structured forms of communication that enable people to reach decisions together. Algorithms would have an important role to play. They would codify what philosopher John O'Neill describes as "rules of thumb, standard procedures, default procedures, and institutional arrangements that can be followed unreflectively and which reduce the scope for explicit judgements," streamlining the planning process so it doesn't become an endless series of meetings. At the same time, we would need some set of rules for how to tie all of the protocols together, and to integrate them with the algorithms, in order to create a unified planning apparatus based on software that is easy to use, transparent in its outcomes, and open to modification.

After all, even if we incorporate qualitative goals into our planning, we still have to solve the socialist calculation problem. Producers still have to make decisions that add up into a coherent production plan.

Freely Associated Producers

Neurath's emphasis on democratic decision making was essential. But by proposing the idea of the protocol, he raised more questions than he could answer, especially with the limited technologies available to him at the time. Towards the end of his life, Neurath spent years trying to determine how semi-literate peasants and urban workers could be incorporated into a planning protocol, via the distribution of simple graphical representations that he called isotypes.

Today, literacy is widespread across the world, and cell phones are common even in remote areas. The possibilities for protocol socialism are correspondingly enlarged. However, true democratic decision-making about production cannot simply be a matter of a perpetual social-media plebiscite scrolling across one's phone screen—for the simple reason that many individuals lack the practical knowledge necessary for making most production decisions.

Participation in making each decision, therefore, generally needs to be limited to those involved in and affected by each decision being made, with only decisions that concern everybody being brought to society as a whole. Coordination should take place, in other words, mostly within and between associations. These associations might be composed of producers, consumers, or other groups of people with common identities and interests.

> *"What would otherwise have been an impossibly long, if not interminable, series of meetings might become, with the help of algorithms and protocols, something more manageable."*

Neurath saw this future dimly, through the lens of the social mobilizations of his time. During World War I, masses of workers joined militant rank-and-file movements demanding workplace democracy, including the Industrial Workers of the World in the US, the Shop Stewards Movement in the UK, the councilists in Germany, and the anarcho-syndicalists in Spain, France, and Italy. An issue that arose in these organizations was how to coordinate production among worker-controlled workplaces. Too often, theorists turned to market prices or price-like labor-time calculations for the answer, anticipating the later Lange-Lerner model of an algorithmic socialism.

Neurath hoped that councils, guilds, and other associations could find another way forward. In particular, he speculated that they might be able to use planning protocols to make their own direct comparisons between different "ways of working"—taking into account many and varied criteria that could not "be reduced to one single unit"—while collaborating with one another to help fulfill society-wide goals.

Today's digital technologies might make it easier for such comparisons and collaborations to occur. The association of pencil producers might be algorithmically assigned tokens or "points"—as in economist Daniel Saros's model of digital socialism—which the association uses to bid on graphite, wood, and other intermediate goods, in an effort to find the best way to make a pencil. Periodically, the association of pencil makers would then meet with other graphite-consuming associations. They would examine existing allocation patterns, consider larger social goals, and alter the graphite allocation protocol accordingly. What would otherwise have been an impossibly long, if not interminable, series of meetings might become, with the help of algorithms and protocols, something more manageable—a streamlined planning process, capable of undertaking complex multi-criteria adjustments.

From any given starting point, the socialists of the future might then begin to alter the overall shape of their productive apparatus. For instance, they might set out to reduce the work week by 10 percent over five years, without a significant loss in productive capacities. Associations of workers and consumers would then consider the options available to them for enhancing productivity levels in the specific areas that concern them. New technologies might improve labor productivity in pencil factories, but require more rapid depletion of forest reserves. Meanwhile, a new way of organizing hospitals might result in less work for nurses, but at the cost of lower-quality elder care.

Where do different associations of workers and consumers stand on these issues?

Associations would make recommendations and reach decisions through the direct comparison of plan options, considering the consequences that each productivity-enhancing innovation would have for other issues that their members care about, such as sustainability and justice. At a certain point, a committee might compare society-wide goals of work reduction to actual achievements, looking at sticking points, theorizing solutions, and adjusting incentives to prioritize certain kinds of labor accordingly.

From this perspective, it is easy to see that a planning process would not emerge fully formed with the push of a button on an algorithmic dashboard. Nor would production be constantly revolutionized—at the cost of dislocating human lives and destroying the environment. Instead, step-by-step adjustments would make the production process ever more rational—in the Neurathian sense, not the capitalist one—across a wide variety of criteria. People themselves would propose, debate, and implement improvements for themselves.

The productive apparatus would have more in common with a "food forest" than a factory—a garden of edible plants, tended for hundreds of years and designed to provide for a multiplicity of needs, spiritual as much as material. It would connect the past to the future, across generations. It would be a common inheritance that made it possible for the masses of humanity to live and work as they wanted. Beyond this shared realm of mutual obligations, an enlarged realm of freedom would progressively open up space for radical experimentation that could be explored by all, without endangering anyone's material security or individual freedom.

A Dance Club for Pencil Makers

Digital technologies will assist in the construction of a socialist society, but the role they will play needs to be clarified. We do not want software to substitute for the price mechanism. No matter how digitally mediated a socialist society becomes, it will never be able to escape the need for democratic deliberation at all levels. Human beings are never simply rule followers. They look beyond the rules, sometimes for social benefit, sometimes for personal advantage, and often for both.

> *"The productive apparatus would have more in common with a 'food forest' than a factory."*

At the same time, we have to accept that deliberating end-lessly is undesirable and doomed to failure. To function at all, a society that replaces the single-minded focus on cost control with multi-criteria decision-making must use algorithms to help clarify the choices to be made and protocols to help structure the way it makes these choices. We cannot rely on a single, unified mechanism for this purpose; we will need many. And open-ended debate must modify these mechanisms when they generate bad results or threaten to give rise to new forms of domination.

In designing our protocols and our algorithms, it is crucial to remember that the point of this process of social transformation is not only to make work better, but also to work less. Too often, socialists have seen work as the highest realization of human freedom. In truth, work will never be an entirely free activity. But in a world no longer beholden to the capitalist growth imperative, advanced technologies can substantially reduce the amount of work demanded of any individual. With greater free time and available space, all individuals will be able to develop their personalities outside of a work-centric identity.

The world's pencil makers would be free to invest themselves in a much wider range of ends, whether starting specialized gyms or dance clubs, joining theatre troupes, or forming amateur scientific societies. A rich and varied life beyond work is only possible if work is organized in a way that is fair, rational, and resistant to whatever forces might emerge to subjugate human beings once again. Instead of waiting for a breakthrough in artificial intelligence to achieve this goal for us, we should begin to develop the protocols of the future today. ⌇⌇⌇

This article could not have been written without the conversation and support of Björn Westergard.

Aaron Benanav is a researcher at Humboldt University of Berlin and the author of *Automation and the Future of Work*.

LOGIC BOOKS

Logic Books is a new series from FSG Originals
and Logic that dissects the way that technology
functions in our everyday lives. The first season,
published in October 2020, consists of four titles
covering a wide range of topics, from the coming
implosion of the ad-driven internet to the central
role of tech in rural China.

To learn more and buy the books, visit

https://logicmag.io/books

Subprime Attention Crisis:
Advertising and the Time
Bomb at the Heart of the
Internet

TIM HWANG

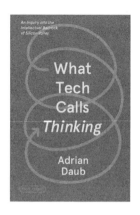

What Tech Calls Thinking:
An Inquiry into the
Intellectual Bedrock of
Silicon Valley

ADRIAN DAUB

Voices from the Valley: Tech
Workers Talk About What
They Do—And How They Do It

BEN TARNOFF
MOIRA WEIGEL

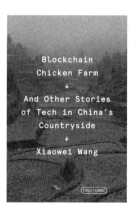

Blockchain Chicken Farm:
And Other Stories of Tech in
China's Countryside

XIAOWEI WANG

[A] well-grounded attack on the fictions that uphold the online advertising ecosystem...One can only hope that this book will be used to pop the bubble that benefits so few.

DANAH BOYD
founder of DATA & SOCIETY

Subprime Attention Crisis

Advertising and the Time Bomb
at the Heart of the Internet

TIM HWANG

(FSGO / *LOGIC*)

Adrian Daub airs out Silicon Valley's smoke and turns over its mirrors in *What Tech Calls Thinking*. This book is a bright, jaunty work of tech criticism.

JOANNE MCNEIL
author of **LURKING**

What Tech Calls Thinking

An Inquiry into the Intellectual Bedrock of Silicon Valley

ADRIAN DAUB

Voices From the Valley gives tech workers—from a cafeteria contractor to a founder who failed up—a new platform to speak for themselves...A glimpse of the business values, politics, motivations and mundanities animating one of the twenty-first century's most opaque and influential power centers. Timely and important.

ANNA WIENER
author of UNCANNY VALLEY

Voices from the Valley

Tech Workers Talk About
What They Do—And How They Do It

BEN TARNOFF, MOIRA WEIGEL

Xiaowei Wang does a tremendous job of weaving together intricate details of China's political and cultural history... *Blockchain Chicken Farm* provides an incisive critique of the possible effects and future of machine learning, artificial intelligence, and automation that narrows the gap between farmlands and the concrete of the cityscape.

DOROTHY R. SANTOS

writer, artist, educator and co-founder of REFRESH

Blockchain Chicken Farm

And Other Stories of Tech
in China's Countryside

XIAOWEI WANG

o hi

We're a small magazine, and we pay our writers.

Writing is hard work, and we believe we can't have a better discourse around technology without compensating the people who are working to improve it.

But we can't do this without your help.

Subscribing is one way to support this project. Making a tax-deductible donation is another.

Logic Magazine is published by the Logic Foundation, a California nonprofit public benefit corporation with 501(c)(3) status.

Your contributions will enable us to keep building a project that's committed to recognizing and rewarding creative labor.

For more information about the Logic Foundation, including how to donate electronically or by check, visit **logicmag.io/donate**.

Other questions? Email **donate@logicmag.io**.

thxbye

L O G I C
upcoming

ISSUE 13: DISTRIBUTION
SPRING 2021

The internet was invented for the purpose of redistribution: to move computing power from one place to another. Today, the cloud both has and has not fulfilled this dream. This issue will explore the distributive aspects of digital technologies: On the one hand, users anywhere can borrow cycles from servers deep in the forests of Oregon or high on the mountains of Guizhou. On the other, those servers belong to just a handful of companies. Data analytics and machine learning have made it possible to optimize supply chains linking every part of the world, but they have not spread production or profits evenly. New futures are always arriving; they are never evenly distributed.

ISSUE 14: KIDS
SUMMER 2021

That is no country for old men. Silicon Valley is notorious for turning *don't trust anyone over thirty* from a counterculture slogan into a business strategy, which can get awkward as its most famous boy geniuses start to go gray. This issue will dig into new and unexpected relationships between youth and technologies. Topics may include: edtech, ageism, and TikTok; privacy protections, parental controls, and cyberbullying; working from home, carework platforms, Mommy blogs, infertility, and online abortion pharmacies; child labor, student debt and youth protest movements; transhumanism and eternal life.

subscribe @ https://logicmag.io